THEORIES OF GRAVITATION:
CERTAIN RAMIFICATIONS

<u>ACKNOWLEDGEMENT</u>

Research output of this scale requires support, encouragement and guidance from everyone besides the intellectual associations. It is my duty to acknowledge their support and express my gratitude for bringing my endeavor to accomplishment.

*At the outset I express my profuse heartfelt thanks to my Research Director **Prof. V. Uma Maheswara Rao**, Department of Applied Mathematics and Registrar, Andhra University, for his constant guidance and invaluable critical insights and encouragement.*

I sincerely thank Prof. D.R.K. Reddy, Former Professor of Applied Mathematics, Dr. M. Vijaya Santhi, Associate Professor and Dr. T. Vinutha, Assistant Professor, Andhra University for their concern for my academic well-being.

I am deeply obliged to Prof. K. Rajendra Prasad, Head, Department of Applied Mathematics, Dr. P. Vijaya Laxmi, Assistant Professor, Andhra University for their cooperation and encouragement.

I thank Dr.K.V.S.Sireesha and Dr.D.Neelima, Assistant Professors, GITAM University, Dr.B.Jagan Mohan Rao, C.R.Reddy College, Dr.D.Ch.Papa Rao, S.K.B.R. College, my co-research scholars, U.Y.Divya Prasanthi, N.Sandhya Rani, Molla Mangesha and especially Y.Aditya for their continual encouragement. I also thank other research scholars of Dept. of Applied Mathematics and my friends for their support and cooperation.

It is time to express my profound sense of gratitude to my beloved parents Smt. G. Tatamma (Late), Sri. Venkateswara Naidu. I am grateful to my wife Bhagya Lakshmi and my daughter Yuktasri whose reinforcement has enabled me to pursue both my professional and academiccareer.

G. Suryanarayana

CONTENTS

Page No.

CHAPTER-1

CHAPTER- 4

FIVE DIMENSIONAL KALUZA-KLEIN AND KANTOWSKI-SACHS PERFECT FLUID MODELS IN $f(R,T)$ GRAVITY

CHAPTER- 5

TILTED BIANCHI TYPE-*I* MODEL IN BRANS-DICKE THEORY AND TILTED BIANCHI TYPE-*II*, *VIII*, *IX* MODELS IN LYRA MANIFOLD

CHAPTER- 6

KANTOWSKI-SACHS TWO FLUID RADIATING MODELS IN BRANS-DICKE THEORY

CHAPTER - 7

KANTOWSKI-SACHS HOLOGRAPHIC MODEL IN SAEZ-BALLESTER THEORY AND KALUZA–KLEIN HOLOGRAPHIC MODEL IN BRANS-DICKE THEORY

166-190

CHAPTER – 8

KANTOWSKI-SACHS DARK ENERGY MODEL IN GENERAL SCALAR-TENSOR THEORY

REPRINTS OF THE PUBLISHED PAPERS

Chapter-1

General Introduction

1.1 Introduction

The thesis titled **"Theories of Gravitation: Certain Ramifications"** consists of eight chapters and deals with the investigation of some spatially homogeneous and anisotropic cosmological models in the frame work of $f(R,T)$ theory of gravity proposed by Harko et al. (2011) and in scalar-tensor theories of gravitation formulated by Sen (1957), Brans-Dicke (1961), Nordtvedt (1970) and Saez-Ballester (1986), which are viable alternatives to Einstein's (1915) theory of gravitation.

Cosmology is the scientific study of the large scale properties of the Universe as a whole. It endeavors to use the scientific method to understand the origin, evolution and ultimate fate of the Universe in which people live in. Similar to any other field of science, cosmology involves the formation of theories or hypotheses about the Universe, which makes specific predictions that, can be tested with observations. It is a known fact that this can be done effectively by constructing the mathematical models (cosmological models), studying their physical behavior and then comparing the same with the present day observations.

Einstein's theory of gravitation has been successful in the study of cosmological models. However, it has been pointed out that this theory lacks certain aspects. Hence, in recent years, several alternative modified theories of gravitation have been proposed to study the Universe through cosmological models. With this motivation, we have taken up the investigation of some spatially homogenous and anisotropic cosmological models in the above mentioned alternative theories of gravitation.

This chapter is organized as follows: section 1.2 is devoted to a brief introduction to Einstein's theory of gravitation. In section 1.3, a brief review of scalar-tensor theories of gravitation, relevant to the proposed study is presented. Section 1.4 deals with review of $f(R,T)$ theory of gravity. In section 1.5, cosmology and cosmological models are discussed briefly. Section 1.6 deals with dark energy. In section 1.7, review of Bianchi space-times and their significance in cosmology is presented. In section 1.8, some aspects of the higher dimensional space-time are presented.

The last section 1.9, describes the problems investigated and the results obtained.

1.2 Einstein's Theory of Gravitation

The general theory of relativity is an astounding accomplishment: Together with quantum field theory, it is now widely considered to be one of the two pillars of modern physics. The theory itself is couched in differential geometry, and is a pioneer for the use of modern mathematics in physical theories, paving way for the gauge theories and string theories that follow. It is no exaggeration to say that general relativity set a new tone for a physical theory, and has truly revolutionized our understanding of the Universe. One of the most striking facts that theory is general remains completely unchanged even after almost an entire century. The field equations that Einstein communicated to the Prussian Academy of Sciences in November 1915 are still the best description of how space-time acts upon on macroscopic scales.

The Einstein's field equations are

$$G_{\mu\nu} = \frac{8\pi G}{c^4} T_{\mu\nu} \qquad (1.2.1)$$

where $G_{\mu\nu}$ is the Einstein tensor, $T_{\mu\nu}$ is the energy momentum tensor, G is Newton's constant, and c is the speed of light. These equations are thought to govern the expansion of the Universe, the behavior of black holes, the propagation of gravitational waves, and the formation of all structures in the Universe from planets to stars, the way up to the clusters and super-clusters of galaxies that we see today. General relativity (G.R.) is thought to be inadequate in the microscopic world of particles and high energies. On all other scales G.R. remains the best.

The great success of general theory, however, has not stopped alternatives being proposed. Even during the very early days after Einstein's publication of his theory; there were proposals being made on how to extend general theory, and incorporate it in a larger and more unified theories. Notable among them are Eddington's theory of connections, Weyl's scale independent theory, and the higher dimensional theories of

Kaluza and Klein. To some extent, these early theories were known to have been influenced Einstein himself.

So, in the recent years, there have been some interesting attempts to generalize the general theory of relativity by incorporating Mach's principle and other certain desired features which are lacking in the original theory. With this objective, various versions of scalar-tensor theories of gravitation have been suggested and have been widely discussed. Among these theories; the scalar-tensor theories of gravitation proposed by Sen (1957) based on Lyra (1951) geometry, Brans and Dicke (1961), Nordtvedt (1970) along with the special case proposed by Schwinger (1970) and Saez and Ballester (1986) is drawing the attention of physicists. Extensive examination of certain aspects of these scalar-tensor theories of gravitation (e.g., exact solutions for field equations, cosmology, the three crucial tests of general relativity etc.) have revealed significant points of similarity and contrast between scalar-tensor theories and Einstein's theory of gravitation. All versions of scalar-tensor theories are based on the introduction of a scalar field variable into the formulation of general relativity and this scalar field together with the metric tensor field g_{ij} forms the scalar-tensor field, representing the gravitational field.

1.3 Scalar-Tensor Theories of Gravitation

Owing to cosmological reasons scalar-tensor gravity was introduced as a definite theory in the early 1960's and inception, has been consistently applied to cosmology. Since then, string theorists using a different approach have found many similarities between string and scalar-tensor theory, prompting further investigation of the latter. Such independent motivations and approaches have spurred the interest of many researchers in scalar-tensor gravity resulting in rapid growth in this area.

1.3.1 Sen theory based on Lyra geometry:

In recent years there has been a lot of interest in alternative theories of gravitation. Noteworthy among them is the theory of gravitation proposed by Sen (1957) based on Lyra (1951) geometry. This geometry is a modified Riemannian geometry in which a gauge function has been introduced into the structure less manifold as a result of which the cosmological constant arises naturally from the geometry. In general relativity, Einstein succeeded in geometrizing gravitation by identifying the metric tensor with gravitational potentials. The scalar field in scalar-tensor theory of Brans-Dicke on the other hand, remains alien to the geometry. Lyra's geometry is continuation of Einstein's principle of geometrisation since both the scalar and tensor fields have more or less intrinsic geometrical significance. Lyra (1951) defined the displacement vector PP^1 between two neighboring points $P(x^i)$ and $P(x^i + dx^i)$ which has the components $\xi^i = x^0 dx^i$ where $x^0(x^i)$ is a non-zero gauge function. The coordinate system x^i together with x^0 form reference system (x^0, x^i). Tensors are characterized by the way in which components transform under a general transformation of coordinates.

The metric

$$ds^2 = g_{ij}x^0 dx^i x^0 dx^j \tag{1.3.1}$$

is an absolute invariant (i.e. invariant under change of reference system).

The components of the affine connection are no longer symmetric in the lower indices and cannot be identified with the Christoffel symbols as is the case in Riemannian geometry. In Lyra geometry the components of the affine connection are not only the functions of Christoffel symbols but also of ϕ_i. The Lyra curvature tensor K^h_{ij}, the contracted curvature tensor K_{ij} and the scalar curvature K can be obtained similar to the Riemannian one.

The scalar constant K is given by

$$K = R(x^0)^{-2} + 3(x^0)^{-1}\phi^i_{;i} + \frac{3}{2}\phi^i\phi_i \tag{1.3.2}$$

where R is the Riemannian curvature scalar, ϕ_i is defined by

$$\phi_i = (x^0)^{-1}\frac{\partial}{\partial x^i}\left\{\log(x^0)^2\right\} . \tag{1.3.3}$$

Choosing the so called normal gauge i.e., $x^0 = 1$ the volume integral in space becomes

$$I = \int k \sqrt{-g}\, d^4 X \qquad (1.3.4)$$

where $k = R + 3\phi^i_{\ ;i} + \dfrac{3}{2}\phi_i \phi^i$.

The field equations in this manifold may be obtained from the variational principle

$$\delta(I + J) = 0 \qquad (1.3.5)$$

where I is given by (1.3.4) and

$$J = \int L \sqrt{-g}\, d^4 X \qquad (1.3.6)$$

here L is the Lagrangian density of matter.

Using the above variational principle, finally Sen (1957) obtained the following field equations

$$R_{ij} - \frac{1}{2} R g_{ij} + \frac{3}{2}\phi_i \phi^i - \frac{3}{4} g_{ij} \phi_l \phi^l = -8\pi T_{ij} \qquad (1.3.7)$$

1.3.2 Brans-Dicke scalar-tensor theory:

Nature's simplest field, the scalar field has been the protagonist of physical theories for a long time. Even before the publication of the theory of general relativity, Nordstrom (1912 1913a, 1913b, 1914) formulated a conformally flat scalar-theory of gravity which was regarded by Einstein as a serious competitor to general relativity for some time (Einstein 1987). Today, a distinguishing feature of modern scalar-tensor cosmology is the gravitational coupling which is time-dependent: this idea dates back to the work of Dirac in 1937 related to the large number hypothesis (Dirac 1937a, b, 1938). Dirac noticed that dimensionless combinations of cosmological constants and fundamental physical constants are connected by a relation that arises naturally provided that one of the "constants" is allowed to change over cosmological timescales. Dirac's choice was to let the gravitational coupling G become time-dependent, keeping the other fundamental constants fixed (Dirac 1938). Next, the development of this idea by Jordan in the following decade took the form of a complete gravitational theory in which G was promoted to do the role of a gravitational scalar field (Jordan 1938, 1952). From an effective point of view, the latter somehow behaves as a form of matter and therefore satisfies a generalized conservation equation built into the theory (Jordan 1938, 1952). Finally, the idea of a scalar-tensor theory reached full maturity with the work of Brans and Dicke (1961). In 1961 they published a new theory that was to become the prototype of the alternative theory to Einstein's general relativity. The theory contains a scalar field which describes gravity together with the metric tensor familiar to general relativity.

The primary motivation for exploring a new gravitational theory comes from cosmology-a new theory explicitly incorporating Mach's principle, which is seen as left out or marginalized in Einstein's theory. The latter contains a few not all aspects of Mach's principle, and also has solutions (plane-fronted waves with parallel rays) that are regarded as explicitly anti-Machian (Ozsvath and Schiicking 1962). The gravitational scalar field is the inverse of the gravitational coupling which is constant in general relativity. In Brans-Dicke theory, however, the gravitational coupling is variable and is determined by all matter in the Universe. A distinguishing feature of the theory is that the cosmological distribution of matter affects local gravitational experiments.

The fact that plausible mechanisms have been discovered at the classical level is another reason for renewed interest in Brans-Dicke and scalar-tensor theories that allow the parameter ω, a variable in scalar-tensor gravity. This was done to assume values of order unity in the early Universe and diverge later there by thus reducing gravity to general relativity at later epochs in the history of the Universe. Finally, renewed interest in scalar-tensor gravity is also due to the use of these theories in inflationary scenarios of the Universe, following the larger and larger consensus obtained by the idea of inflation after Guth's and Linde's seminal papers in the early 1980's. As the attractor solution of general relativity in phase space the de Sitter exponentially expanding Universe fails to provide a graceful exit from an inflationary epoch, its analogue in Brans-Dicke theory a power-law attractor solution does the job. More recently, the astonishing discovery that the Universe is undergoing at an accelerated expansion today. Approximately 70% of the energy of the Universe is in a form with negative pressure (dark energy), this spurred interest in scalar-tensor theories which account for such features with a long-range gravitational scalar field.

In the hope of extending general relativity by incorporating Mach's principle, Brans and Dicke (1961) have proposed a theory consisting a long range scalar field interacting equally with all forms of matter (with the exception of electromagnetism). They noted, following Dirac (1938) and Sciama (1959), that the Newtonian gravitational constant G is related to the mass M and radius R of the visible Universe by

$$G \sim \frac{Rc^2}{M} \tag{1.3.8}$$

(The numbers are approximate). This suggests that G is a (scalar) function determined by the matter distribution. Their theory is formally equivalent to the one previously considered by Jordan (1955).

In order to generalize the equations of general relativity, Brans and Dicke (1961) formulated their variational principle, which differs from that of general relativity, namely,

$$\delta \int [R + (16G)L]\sqrt{-g}\, d^4 x = 0 \tag{1.3.9}$$

In that G is replaced by ϕ^{-1} which now comes inside the action integral. There are also additional terms as scalar nature of ϕ such as to be take its account

$$\delta \int \left[\phi R + 16GL - \frac{\omega \phi_{,i} \phi^{,i}}{\phi} \right] \sqrt{-g}\, d^4 x = 0 . \tag{1.3.10}$$

where ϕ is the scalar field, R is the usual scalar curvature, L is a function of matter variables and metric tensor components (not of scalar field ϕ) and ω is a dimensionless constant.

The field equations obtained by the variation of g_{ij} and ϕ take the form

$$R_{ij} - \frac{1}{2} R g_{ij} = -8\pi \phi^{-1} T_{ij} - \omega \phi^{-2} \left(\phi_{,i} \phi_{,j} - \frac{1}{2} g_{ij} \phi_{,r} \phi^{,r} \right)$$
$$- \phi^{-1} (\phi_{i;j} - g_{ij}\, \phi_{;r}{}^{,r}) \tag{1.3.11}$$

and $\phi_{,r}{}^{,r} = 8\pi(3+2\omega)^{-1}T$ $\qquad$ (1.3.12)

where $T = g^{ij}T_{ij}$. Here the metric has signature +2, a comma denotes partial differentiation, a semi colon denotes covariant differentiation and the velocity of light 'c' is taken to be unity. The main difference between the Brans-Dicke theory and Einstein theory lies in the gravitational field equations, which determine the metric field g_{ij}, rather than in the equations of motion. The energy momentum tensor of matter T_{ij} satisfies the local matter-energy conservation law;

$$T^{ij}{}_{;j} = 0 \qquad (1.3.13)$$

which also represents equation of motion and is a consequence of the field equations (1.3.11) and (1.3.12).

A comparison of the above equations with Einstein's equations shows that the Brans-Dicke theory goes over to general relativity in the limit

$$\omega \to \infty, \phi = \text{constant} = G^{-1}.$$

The above modification of Einstein's theory involves violation of "strong principle of equivalence" on which Einstein's theory is based.

But this does not violate "weak principle of equivalence" for example, the paths of test particles in a gravitational field are still independent of their masses. Thus, Brans-Dicke theory now can be described as a theory for which the gravitational force on an object is partially due to the interaction with a scalar field, and partially due to a tensor interaction.

Further discussions, by Brans and Dicke (1961), about the field equations in (1.3.11) and (1.3.12) have included and analysis of the weak field equations, study of the three standard tests, comparison with the work of Jordan (1955), discussions of boundary conditions for ϕ, investigations of cosmology and the general relationship to Mach's principle.

At present there is no evidence to preclude the validity of the Brans-Dicke scalar-tensor theory, this theory doesn't predict an anomalous gravitational red shift, it gives values for the gravitational deflection of light rays and the perihelion advance of planetary orbits different from those of Einstein's theory (Dicke 1964, Brans-Dicke 1961). But in view of the relatively large discrepancies in the measurements of the deflection of star light near the sun's limb during a total eclipse (Dicke 1967) and the measurements of the oblateness of the sun (Dicke and Goldenberg 1967), it is concluded that Brans-Dicke theory is not a conflict with observations, provided that $\omega \geq 6$.

Very recently, this theory has been applied to more interesting problems in astrophysics in order to appreciate the implications of the addition of a long range scalar-interaction. By comparing the predictions of this theory with those of Einstein's theory, one may hope to obtain important differences which might be used to decide between the two theories. For example, while Solmona (1967) has shown that certain gross features of a cold neutron star remain unchanged by the presence of the strength of the scalar field, Morganstern and Chiu (1967) have shown that if a neutron star is observed to exhibit symmetric radial pulsation, then the existence of the scalar field may be ruled out.

As a consequence of the recent lunar ranging experiments (Williams et al 1976, Shapiro et al 1976) one can conclude that Brans-Dicke parameter $|\omega| \geq 500$. It has been pointed out that there is no theoretical reason to restrict ω to positive values (Smalley and Eby 1976). In view of this one might as well conclude that Brans-Dicke theory with some large values of $|\omega|$ is a correct theory.

In recent years, Rao et al. (2009) have obtained axially symmetric string cosmological models in Brans-Dicke theory of gravitation. Chen and Wu (2011) have discussed cosmological constraint on Brans-Dicke theory. Zeyauddin and Shri Ram (2011) have studied spatially homogeneous Bianchi type-V cosmological models in Brans-Dicke theory. Kucukakca et al. (2012) have discussed LRS Bianchi type-I Universes exhibiting No ether symmetry in the Brans-Dicke scalar–tensor theory. Sharif and Waheed (2013) have investigated cosmology of some holographic dark energy models in chameleonic Brans-Dicke gravity. Oikonomou and Karagiannakis (2014) have studied antigravity in

$f(R)$ and Brans-Dicke theories. Rao and Jayasudha (2015a) have studied Bianchi type-V dark energy model in Brans-Dicke theory. Rao et al. (2015a) have discussed five dimensional FRW radiating models in Brans-Dicke theory of gravitation. Rao and Suryanarayana (2015a) have studied a tilted cosmological model in Brans-Dicke theory of gravitation.

1.3.3 Nordtvedt general scalar-tensor theory:

Brans and Dicke have proposed a modification of Einstein's theory of gravitation through the introduction of a scalar field ϕ in the field equations to make things more consistent with Mach's principle and less reliant on the absolute properties of space. But in view of the recent experimental evidence it is argued that if the Brans-Dicke theory of gravitation is a correct theory, than the value of the parameter ω in this theory has to be as large as, or even greater that of 40,000. With such a large value of ω, it is difficult to distinguish between the Brans-Dicke theory of gravitation and the general theory of relativity, at least from their consequences. On the other hand, there is no a priori reason to exclude the introduction of any long range scalar field in the evolution of the Universe which might be quite important at some epoch, one may explore the possibility of a general scalar-tensor theory with ω as a time-dependent function. Nordtvedt modified the Brans-Dicke theory where ω now becomes a function of scalar field ϕ instead of being a constant. Within the frame work of Nordtvedt's general scalar-tensor theory Barker proposed a particular $\omega - \phi$ relation, where the local gravitational constant in the Newtonian approximation does not change with time. Also, Schwinger (1970) & Kimball and Yee (1974) have formulated a scalar-tensor theory as a mass-varying theory, that can be put in the form of a standard scalar- tensor theory with a suitable choice of the function $\omega(\phi)$ and can be carried out after a transformation to "particle units".

The field equations of this general class of scalar-tensor gravitational theories are obtained from the variational principle

$$\delta \int \left[16\pi L + \phi R + \frac{\omega(\phi)}{\phi} \phi_{,i}\phi^{,i} \right] \sqrt{-g}\, d^4 x = 0. \qquad (1.3.14)$$

In equation (1.3.14) L is the matter Lagrangian which could include electromagnetic fields and other forces also. Setting the arbitrary function of the scalar field equal to $\dfrac{\omega(\phi)}{\phi}$ is purely notational to facilitate comparison of this theory with that of Brans-Dicke theory in which

$$\omega(\phi) = \text{constant.} \tag{1.3.15}$$

Applying the variation $\delta\phi$ and δg_{ij} to the integral (1.3.14), one obtains the field equations for the scalar and tensor fields;

$$\phi_{;r}^{;r} = \frac{8\pi T}{3+2\omega} - \frac{1}{(3+2\omega)}\frac{d\omega}{d\phi}\phi_{,i}\phi^{,i} \tag{1.3.16}$$

And

$$R_{ij} - \frac{1}{2}Rg_{ij} = -8\pi\phi^{-1}T_{ij} - \omega\phi^{-2}\left(\phi_{,i}\phi_{,j} - \frac{1}{2}g_{ij}\phi_{,r}\phi^{,r}\right)$$
$$- \phi^{-1}(\phi_{i;j} - g_{ij}\,\phi_{;r}^{,r}) \tag{1.3.17}$$

Here T_{ij} is the stress-energy tensor of the matter defined by

$$T^{ij} = 2(-g)^{\frac{-1}{2}}\frac{\partial}{\partial g_{ij}}\left((-g)^{\frac{1}{2}}L\right), \tag{1.3.18}$$

with $T^{ij}{}_{;j} = 0.$ \hfill (1.3.19)

This general class of scalar-tensor theories would seem to lead to a super richness (or arbitrariness) of possible theories. Considering the static spherically symmetric solution for a point mass source, Nordtvedt (1970) found a variety of experimental consequences of $\omega' = d\omega/d\phi \neq 0,$ including a contribution to the rate of precession of Mercury's perihelion. If experiments suggest that the more general scalar-tensor theory $(\omega' \neq 0)$ is valid, then we are in need of some physical principle which will suggest a particular form for out of all possible functions.

Schwinger (1970) has formulated a scalar-tensor theory as a mass varying theory by modifying the Einstein Lagrange function by adding a scalar field in such a way so as to render the theory conformally invariant. But it can be put in the form of Nordtvedt's theory with a choice of $\omega(\phi)$ as

$$3 + 2\omega(\phi) = \frac{1}{\lambda\phi}, \quad \lambda = \text{constant}.$$

A general class of scalar-tensor theory of gravitation, which may be said to be a generalization of the Brans-Dicke theory, has been discussed by many authors. Bergmann (1968) has discussed the effect of momenta and mass of particles in a generalized scalar-tensor theory. Wagoner (1970) has presented an analysis of general scalar-tensor theory of gravitation containing two arbitrary functions of scalar field. Barker (1978), Banerjee and Duttachoudary (1980), Banerjee and Santos (1981a, b), Rao and Reddy (1982), Van den Bergh (1982, 1983), Shanti and Rao (1989, 1990a, b) are some of the authors who have investigated several aspects in this theory.

1.3.4 Saez-Ballester scalar-tensor theory:

Saez and Ballester (1986) developed a scalar-tensor theory in which the metric is coupled with a dimension less scalar field. This coupling gives a satisfactory description of weak fields. Inspite of the dimensionless character of the scalar field, an antigravity regime appears. This theory suggests a positive way to solve the missing matter problem in non-flat FRW cosmologies. Saez and Ballester (1986) assumed the Lagrangian

$$L = R - \omega \phi^n \left(\phi_{,\alpha} \phi^{,\alpha} \right) \tag{1.3.20}$$

where R is the curvature, ϕ is the dimensionless scalar field, ω and n are arbitrary dimensionless constants and $\phi^{,i} = g^{ij}\phi_{,j}$. For scalar field having the dimension $\phi = G^{-1}$, the Lagrangian given by equation (1.3.20) has different dimensions. However, it is a suitable Lagrangian in the cases of a dimensionless scalar field. From the Lagrangian one can build the action

$$I = \int_{\Sigma}(L + GL_m)\sqrt{-g}\; dx\,dy\,dz\,dt \tag{1.3.21}$$

where L_m is the matter Lagrangian, $g = |g_{ij}|$, Σ is an arbitrary region of integration and $G = -8\pi$. By considering arbitrary independent variations of the metric and the scalar field vanishing at the boundary of Σ, the variational principle

$$\delta I = 0 \tag{1.3.22}$$

leads to the Saez-Ballester (1986) field equations for combined scalar and tensor fields given by

$$R_{ij} - \frac{1}{2}Rg_{ij} - \omega\phi^n\left(\phi_{,i}\phi_{,j} - \frac{1}{2}g_{ij}\phi_{,r}\phi^{,r}\right) = -8\pi T_{ij} \tag{1.3.23}$$

and the scalar field ϕ satisfies the equation

$$2\phi^n \phi^{,i}_{;i} + n\phi^{n-1}\phi_{,r}\phi^{,r} = 0 \qquad\qquad (1.3.24)$$

Also $\quad T^{ij}_{;j} = 0$ $\qquad\qquad\qquad\qquad\qquad\qquad\qquad$ (1.3.25)

This is a consequence of the field equations (1.3.23) and (1.3.24).

Saez (1987) discussed the initial singularity and inflationary Universe in this theory. He has also obtained non-singular FRW model in the case $k=0$. The study of cosmological models in the framework of scalar-tensor theories has been the active area of research for the last few decades. In particular, Singh and Agrawal (1991), Shri Ram and Tiwari (1998), Reddy and Naidu (2007) and Rao et al. (2007), Shri Ram et al. (2008) are some of the authors who have investigated several aspects of the cosmological models in Saez-Ballester scalar-tensor theory. In recent years, Rao et al. (2011) have discussed anisotropic Universe with cosmic strings and bulk viscosity in this scalar-tensor theory of gravitation. Shri Ram et al. have studied (2011) Bianchi type-V viscous fluid cosmological models in Saez-Ballester theory of gravitation. Jamil et al. (2012) have obtained Bianchi type-I cosmology in generalized Saez-Ballester theory via Noether gauge symmetry. Pradhan et al. (2013) have studied accelerating Bianchi type-V cosmology with perfect fluid and heat flow in Saez-Ballester theory. Reddy et al. (2014a) have investigated Bianchi type-V bulk viscous string cosmological model in Saez-Ballester scalar-tensor theory of gravitation. Chandel and Shri Ram (2015) have discussed anisotropic cosmological models with bulk viscosity and particle creation in Saez-Ballester theory of gravitation.

1.4 $f(R,T)$ Modified Theory of Gravity

The limits of general relativity become the focal points with the emergence of the "dark Universe" scenario. For almost thirty years there is a concept that, if gravity is governed by Einstein's field equations, there should be a substantial amount of "dark matter" in galaxies and clusters. More recently, "dark energy" has also been found to be required in order to explain the apparent expansion of the Universe at an accelerating rate. Indeed, if general relativity is correct, it now seems that around 96% of the Universe should be in the form of energy densities that do not interact electromagnetically. Such an odd composition, favored at such high confidence, has led some to speculate on the possibility that general relativity may not, in fact, be the correct theory of gravity to describe the Universe on the larger scales. The dark Universe may be just another signal that we need to go beyond Einstein's theory.

There are two representative directions to address the issue of cosmic acceleration. One of these is to introduce the "exotic energy component" in the context of general relativity. Several candidates have been proposed (Sahni and Starobinsky 2000; Kamenshchik et al. 2001; Bento et al. 2002; Padmanabhan 2002; Sen 2002; Caldwell 2002; Nojiri and Odintsov 2003a; Sahni 2004; Feng et al. 2005; Guo et al. 2005; Padmanabhan 2008) in this perspective to explore the nature of dark energy. The other direction is to modify the Einstein Lagrangian i.e., modified gravity theory such as f(R) gravity (Nojiri and Odintsov 2007; Sotiriou and Faraoni 2010). Recently, Harko et al. (2011) generalized $f(R)$ gravity by introducing an arbitrary function of the Ricci scalar R and the trace of the energy-momentum tensor T. The dependence of T may be introduced by exotic imperfect fluids or quantum effects (conformal anomaly). As a result of coupling between matter and geometry motion of test particles is non-geodesic and an extra acceleration is always present. In $f(R,T)$ gravity, cosmic acceleration may result not only due to geometrical contribution to the total cosmic energy density but it also depends on matter contents. This theory can be applied to explore several issues of current interest and may lead to some major differences.

Harko et al. (2011) developed $f(R,T)$ theory of gravity, where the gravitational Lagrangian is given by an arbitrary function of the Ricci scalar R and of the trace T of the stress-energy tensor. They have obtained the gravitational field equations in the metric formalism, as well as, the equations of motion for test particles, which follow from the covariant divergence of the stress-energy tensor. The $f(R,T)$ gravity model depends on a source term, representing the variation of the matter stress energy tensor with respect to the metric. In $f(R,T)$ gravity, the field equations are obtained from the Hilbert-Einstein type variation principle.

The action principle for this modified theory of gravity is given by

$$S = \frac{1}{16\pi G}\int f(R,T)\sqrt{-g}\,d^4x + \int L_m\sqrt{-g}\,d^4x \tag{1.4.1}$$

where $f(R,T)$ is an arbitrary function of the Ricci scalar R and of the trace T of the stress energy tensor of matter and L_m is the matter Lagrangian. The stress energy tensor of matter is

$$T_{ij} = -\frac{2}{\sqrt{-g}}\frac{\partial(\sqrt{-g})}{\partial g^{ij}}L_m, \; \Theta_{ij} = -2T_{ij} - pg_{ij}. \tag{1.4.2}$$

Using gravitational units (by taking G as unity) the corresponding field equations of $f(R,T)$ gravity are obtained by varying the action principle (1.4.1) with respect to g_{ij} as

$$f_R(R,T)R_{ij} - \frac{1}{2}f(R,T)g_{ij} + (g_{ij}\nabla^i\nabla_i - \nabla_i\nabla_j)f_R(R,T) =$$
$$8\pi T_{ij} - f_T(R,T)T_{ij} - f_T(R,T)\Theta_{ij} \tag{1.4.3}$$

where $f_R = \dfrac{\delta f(R,T)}{\delta R}, f_T = \dfrac{\delta f(R,T)}{\delta T} \,\&\, \Theta_{ij} = g^{\alpha\beta}\dfrac{\delta T_{\alpha\beta}}{\delta g^{ij}}.$

Here ∇_i is the covariant derivative and T_{ij} is usual matter energy-momentum tensor derived from the Lagrangian L_m.

It can be observed that when $f(R,T) = f(R)$, then (1.4.3) reduce to field equations of $f(R)$ gravity.

It is mentioned here that these field equations depend on the physical nature of the matter field. Many theoretical models corresponding to different matter contributions for $f(R,T)$ gravity are possible. However, Harko et al. (2011) gave three classes of these models

$$f(R,T) = \begin{cases} R + 2f(T), \\ f_1(R) + f_2(T), \\ f_1(R) + f_2(R)f_3(T). \end{cases}$$

(1.4.4)

Harko et al. (2011) have discussed FRW cosmological models in this theory by choosing appropriate function $f(T)$.

Jamil et al. (2012) have reconstructed some cosmological model in $f(R,T)$ theory of gravity. Reddy et al. (2013a) have discussed a Bianchi type-*III* cosmological model in $f(R,T)$ gravity in presence of perfect fluid. Ahmad and Pradhan (2014) have discussed Bianchi type-*V* Cosmology in $f(R,T)$ gravity with $\Lambda(T)$. Yadav (2014) has studied Bianchi type-*V* string cosmological model with the help of power law in $f(R,T)$ gravity. Mahanta (2014) has constructed Bulk viscous cosmological models in $f(R,T)$ theory of gravity. Recently, Sahoo et al. (2015) have investigated Kaluza-Klein cosmological model with a particular choice of functional as $f(R,T) = f(R) + f(T)$ in $f(R,T)$ gravity.

1.5 Cosmology and Cosmological Models

The word "Cosmology" is derived from Greek word "Kosmos", which means the beauty of the sky. Cosmology is a branch of science that deals with the study of large scale structure of the Universe. The Universe consists of stars, star clusters and galaxies or the nebulae, pulsars, quasars as well as cosmic rays and background radiation. The basic problem in cosmology is the dynamics of the system. The force of gravity is the fundamental force that keeps solar system, stars and galaxies together. The other long range interactions such as electromagnetic force may be disregarded because the galaxies which are the major constituents of the Universe as well as the intergalactic medium are known to be electrically neutral.

It is well known that Einstein's general theory of relativity is a satisfactory theory of gravitation correctly predicting the motion of test particles and photons in curved space-time; but in order to apply the same to the Universe; one has to introduce simplifying assumptions and approximations.

The first approximation that is usually made is that of continuous matter distribution. The most powerful assumption in standard cosmological theory is that of homogeneity and isotropy, often referred to as the "cosmological principle". According to the ordinary cosmological principle, a typical fundamental observer cannot tell where he is located or in which direction he is looking by large scale observations. This implies that on a sufficiently large scale, the Universe has no preferred position or direction. The perfect cosmological principle states that besides the implications of the ordinary cosmological principle the Universe also looks the same to a fundamental observer at all epochs. Physically, this implies that there is no preferred position, preferred direction or preferred epoch in the Universe. Thus by invoking the cosmological principle, we, grossly, idealize the

Universe and model it by a simple macroscopic perfect fluid (devoid of shear-viscous, bulk-viscous and heat conductive properties).

Its energy momentum tensor T_{ij} is given by

$$T_{ij} = (\rho + p)u_i u_j + p g_{ij} \tag{1.5.1}$$

where ρ is its proper density, p is its pressure and u_i is the four-velocity vector of the fluid particles (stars etc).

The aim of cosmology is to determine the large scale structure of the physical Universe. Cosmologists construct mathematical models which they believe represent the Universe as a whole concentrating on its large scale features. They compare these models with the Universe as observed by the astronomers.

The cosmological principle allows us to describe the most general homogeneous and isotropic space time by the Robertson-Walker metric

$$ds^2 = dt^2 - R^2(t)\left[\frac{dr^2}{(1-kr^2)} + r^2 d\theta^2 + r^2 \sin^2\theta\, d\varphi^2\right] \tag{1.5.2}$$

where $R(t)$ is the cosmic scale factor, k is a constant which by a suitable choice of r can be chosen to have the values +1, 0 or -1 according as the Universe is compact, flat or non compact respectively. The coordinates (r,θ,φ,t) form a commoving coordinate system in the sense that the cosmic fluid is at rest with respect to the coordinate system.

Friedmann (1922) was the first to investigate the evolution of the function $R(t)$ using Einstein field equations for all three curvatures. It has been, both experimentally and theoretically, established that the present Universe is spatially homogeneous and isotropic and therefore can be described by the Friedmann-Robertson-Walker (FRW) model (Partridge and Wilkinson 1967, Ehlers et al. 1968). However, there's recent evidence for a small amount of anisotropic (Boughn et al. 1981) and a small magnetic field over cosmic distant scale (Sofue et al. 1979). Although, this evidence is not conclusive, the mere possibility of even small anisotropy and a small magnetic field at the present time would suggest very large departure from (FRW) model at early stages of evolution of the Universe. Thus, it is useful to study cosmological models which may be highly anisotropic and may contain dynamically important electromagnetic fields. For the sake of simplicity it is useful to restrict oneself to models that are spatially homogeneous. The spatially homogeneous and anisotropic Bianchi models present a medium between Friedmann-Robertson-Walker models that are completely inhomogeneous and anisotropic Universes and thus play an important role in current modern cosmology.

1.6 Dark Energy

During the last decades, many independent observations (Riess et al., 1998; Perlmutter et al., 1999; Bennett et al., 2003; Tegmark et al., 2004a; Allen et al., 2004) have provided growing evidence that the Universe is filled with two "dark" ingredients, a collision less component, able to cluster on sub-horizon scales, called dark matter and an almost uniform component with negative pressure, called **dark energy**, whose physical nature is still largely unknown. The dark matter component yields almost one-quarter of the total cosmic energy today, while the dark energy is responsible for about 70%. The sum of their contribution to the present-day cosmic energy budget is just impressive: around

96% of the total. The visible material, has drawn the attention of all physicists and astronomers for a millennia, now appears as a sort of minor "detail" in the cosmos.

The discovery that almost three-quarters of the present cosmic energy density is ascribed to be an almost uniform dark energy component that is able to produce, via its negative isotropic pressure, the accelerated expansion of the Universe, represents the most severe crisis of contemporary physics. At the same time, this discovery opens the door to new theoretical speculations and represents the new frontier for observational cosmology, a joint effort to constrain the dynamical properties of the dark cosmic components and to unveil their physical nature. In this sense, the discovery of the "dark side" of the Universe represents a formidable challenge for the cosmological research of the 21st century.

It is well-known that the simplest and elegant way to explain accelerating expansion of the Universe is the inclusion of Einstein's cosmological constant (Sahni and Starobinsky 2000; Peebles and Ratra 2003), however, the two deep theoretical problems (namely the "fine-tuning" and the "coincidence" one) led to the notion of "dark energy". The dynamical nature (i.e. composition and origin) of dark energy, at least in an effective level, can arise from various scalar fields such as a canonical scalar field (quintessence) (Ratra and Peebles 1988; Wetterich 1988; Liddle and Scherrer 1999), a phantom field (Caldwell 2002; Caldwell et al., 2003), that is a scalar field with a negative sign of the kinetic term, or the combination of quintessence and phantom in a unified model named quintom (Feng et al. 2005; Guo et al. 2005; Cai et al. 2010).

In particular, a model named holographic dark energy has been discussed extensively (Cohen et al. 1999; Hsu 2004; Li 2004; Huang and Li 2005). The model is constructed by considering the holographic principle and some features of the theory of quantum gravity. According to the holographic principle, the number of degrees of freedom in a bounded system should be finite and corresponds to the area of its boundary. By applying the principle to cosmology, one can obtain the upper bound of the entropy contained in the Universe. For a system with size L and UV cut-off Λ without decaying into a black hole, it is required that the total energy in a region of size L should not exceed the mass of a black hole of the same size; thus $L^3\rho_\Lambda \leq LM^2_P L$. The largest L allowed is the one saturating this inequality, thus $\rho_\Lambda = 3c^2 M_p^2 L^{-2}$, where c is a numerical constant and M_p is the reduced Planck mass $M_p^2 = 1/8\pi G$. It just means that there is duality between UV cut-off and IR cut-off. The UV cut-off is related to the vacuum energy, and the IR cut-off is related to the large scale of the Universe, for example the Hubble horizon, event horizon or particle horizon (Cohen et al. 1999; Hsu 2004; Li 2004), and even with an arbitrary cut-off (Horvat 2004) as discussed. Recently, Aghamohammadi et al. (2011), Shen and Zhao (2012), Malekjani et al. (2013), Rahman and Ansari (2014) and Adhav et al. (2015) have investigated different aspects of holographic dark energy.

1.7 Bianchi Space-Times

Bianchi type models are spatially homogeneous and anisotropic Universe models. These models are nine in number, but their classification permits to split them in two classes. There are six models in class A (I, II, VI_1, VII, VIII and IX) and five in class B (III, IV, V, VI_h and VII_h). Spatially homogeneous cosmological models play an important role in understanding the structure and properties of the space of all cosmological solutions of Einstein field equations. These spatially homogeneous and anisotropic models are the exact solutions of Einstein Field equations and are more general when compared to Friedman models in the sense that they can provide interesting results pertaining to the anisotropy of the Universe. It is worth mentioning that, the issue of global anisotropy has gained a lot of research interest in recent times. The standard cosmological model (ΛCDM) based upon the spatial isotropy and flatness of the Universe is consistent with the data from precise measurements of the CMB temperature anisotropy (Hinshaw et al. 2009) from Wilkinson Microwave Anisotropy Probe (WMAP). However, the ΛCDM model suffers from some anomalous features at large scale and signals a deviation from the usual geometry of the Universe.

Recently released Planck data (Ade et al. 2014) show a slight red shift of the primordial power spectrum from the exact scale invariance. It is clear from the Planck data that, ΛCDM model does not fit in well to the temperature power spectrum at low multipoles. Also, precise measurements from WMAP predict asymmetric expansion with one direction expanding differently from the other two transverse directions at equatorial plane (Buiny et al. 2006) which signals a non trivial topology of the large scale geometry of the Universe (Watanabe et al. 2009; Tripathy 2014). Using Einstein definition, Banerjee and Sen (1997) studied energy distribution with Bianchi type I space-time.

The Bianchi-type space-times are generally defined by the following metric:

$$ds^2 = -dt^2 + dl^2, \quad \text{where } dl^2 = g_{ab}dx^a dx^b.$$

Here dl^2 is the three-dimensional line element. According to MacCallum (1979); "spatially homogeneous cosmological models play an important role in attempts to understand the structure and properties of the space of all cosmological solutions of Einstein field equations".

1.8 Higher Dimensional Space-Times

The first systematic studies of higher dimensional geometry date back to the likes of Riemann, Cayley and Grassmann in the mid nineteenth century. It lies at the heart of general relativity, where space and time form part of a curved 3+1 dimensional manifold. Of course, Riemannian geometry is not restricted to 3+1 dimensions, so we have the tools to study gravitational theories in higher dimensions. Indeed, this is more than just a theoretical curiosity. Superstring theory, arguably our best candidate for a quantum theory of gravity, can only be formulated consistently in 10 dimensions.

The problem now is a phenomenological one: Gravity does not behave like a 10 dimensional force in our experiments and observations. Perhaps the simplest observation along these lines is the stability of earth's orbit. In D dimensions of space-time, the Newtonian potential due to a point source will typically go like $1/r^{D-3}$. For $D \neq 4$, it follows that we cannot have stable planetary orbits, and so it is clear that gravity should not appear 10 dimensional on solar system scales. We use the word appear, because there exist gravitational models where the extra dimensions are hidden from experiment, but which open up at shorter and/or larger distances.

So here, we will present a higher dimensional Kaluza-Klein gravity that has been proposed. Kaluza-Klein theory grew out of an attempt to unify gravity and electrodynamics (Nordstrom 1914; Kaluza 1921; Klein 1926a, b). The basic idea was to consider general relativity on a 4+1 dimensional manifold where one of the spatial dimensions was taken to be small and compact. One can perform a harmonic expansion of all fields along the extra dimension, and compute an effective 3+1 dimensional theory by integrating out the heavy modes. This idea has been embraced by string theorists who compactify 10 dimensional string theories and 11 dimensional super gravity / M-theory on compact manifolds of 6 or 7 dimensions respectively, often switching on fluxes and wrapping branes on the compact space (Grana 2006). Each different compactification gives a different effective 4-

dimensional theory, so much so that we now talk about an entire landscape of effective theories (Susskind 2003).

Assuming that the extra dimensions have been stabilized, the late-time dynamics of Kaluza-Klein theories are most easily understood at the level of the 4D effective theory. We will show, this generically corresponds to a 4D gravity theory with extra fields (Clifton et al. 2012). In early times, when the 3 dimensional spaces were comparable in size to the extra dimensions but later, the effective description clearly broke down. This formed the basis of Kaluza-Klein cosmology where one can ask a profound question as to why and how the 3 extended dimensions of space were able to grow large, while the extra dimensions remained microscopically small. It seems quite obvious to say that a fully satisfactory answer to this question has yet to emerge.

1.9 Problems Investigated

In chapter-2, spatially homogeneous and anisotropic Bianchi type-*III*, *V* and *VI₀* cosmological models filled with perfect fluid in the framework of $f(R,T)$ gravity proposed by Harko et al. (2011)

have been studied with an appropriate choice of the function $f(R,T)$. Some important features of the models, thus obtained, have been discussed. It is observed that all the anisotropic models are expanding, non-rotating, accelerating and are also free from singularities.

Chapter-3 deals with the investigations of a spatially homogeneous and anisotropic Bianchi type-*II*, *VIII* and *IX* cosmological models filled with perfect fluid in the framework of $f(R,T)$ gravity with a suitable choice of a function $f(R,T)$. More over some important features the same are also discussed. The models obtained and presented here are non-singular, anisotropic, expanding, non-rotating and accelerating.

In chapter-4, we have obtained spatially homogeneous, anisotropic five dimensional Kaluza-Klein and Kantowski-Sachs cosmological models in the presence of perfect fluid source in the framework of $f(R,T)$ gravity proposed by Harko et al. (2011) with a general choice of the function $f(R,T) = f_1(R) + f_2(T)$. Field equations are obtained in presence of perfect fluid for both models and exact solutions of field equations thus obtained are presented. It is observed that both the models are singularity free.

Chapter-5 is devoted to the study of titled Bianchi type-I cosmological model in Brans-Dicke scalar-tensor theory of gravitation and tilted Bianchi type-II, VIII and IX cosmological models in the frame work of a scalar-tensor theory proposed by Sen (1957) based on Lyra (1951) geometry in the presence of a perfect fluid source. To obtain a solution in both the models, it is assumed that expansion scalar is proportional to shear scalar i.e. $\sigma^2 \propto \theta$, which leads to a relation between metric potentials. Some physical and geometrical properties of the model are also discussed in the study.

In chapter-6, an attempt has been made to investigate the Kantowski-Sachs radiating cosmological models filled with two fluids source in a scalar-tensor theory of gravitation as proposed by Brans and Dicke (1961). Here we consider both the cases, when the dark energy is minimally coupled with barotropic fluid and its direct interaction with it. Some important features of the models, thus obtained, have been also discussed. In these two cases equation of state (EoS) parameter of dark energy is in agreement with the observational data.

Chapter-7 is incorporated with the Kantowski-Sachs holographic cosmological model in the frame work of Saez-Ballester (1986) theory and Kaluza-Klein holographic cosmological model in the frame work of Brans-Dicke (1961) theory. To acquire a determinate solution of the field equations in both the models we have used the scalar expansion that is proportional to the shear scalar. The chapter throws light on physical properties and astrophysical parameters of the models. These minimally interacting models are useful to describe the behavior of dark energy.

Chapter-8 elaborates spatially homogeneous Kantowski-Sachs dark energy cosmological models in Nordtvedt (1970) general scalar- tensor theory of gravitation with the help of a special case proposed by Schwinger (1970). Anisotropic as well as isotropic exact models thus obtained are free from singularities, apart from this some important features of the models have also been discussed. These exact models represent not only the early stages of evolution but also the present Universe.

Chapter-2

Bianchi type-*III*, *V* and *VI₀* perfect fluid models in *f(R,T)* gravity

* Work presented, in this chapter is covered by the following publication:

➢ **Bianchi Type *III*, *V* & *VI₀* Universes in *f(R, T)* gravity: The African Review of Physics, 10, 0017 (2015).**

2.1 Introduction

In recent years, there has been a lot of interest in alternative theories of gravitation. In view of the late time acceleration of the Universe, the existence of the dark matter and dark energy, very recently, modified theories of gravity have been developed. Harko et al. (2011) developed $f(R,T)$ theory of gravity, where the gravitational Lagrangian is given by an arbitrary function of the Ricci scalar R and of the trace T of the stress-energy tensor. They have obtained the gravitational field equations in the metric formalism, as well as, the equations of motion for test particles, which follow from the covariant divergence of the stress-energy tensor. Some particular models corresponding to specific choices of the function $f(R,T)$ are also presented; they have also demonstrated the possibility of reconstruction of arbitrary FRW cosmologies by an appropriate choice of a function $f(T)$. In the present model the covariant divergence of the stress energy tensor is non-zero. Hence the motion of test particles is non-geodesic and an extra acceleration due to the coupling between matter and geometry is always present.

Sharif and Zubair (2012a) have investigated anisotropic Universe models with perfect fluid and scalar field in $f(R,T)$ gravity. Sharif and Zubair (2013a) have discussed energy conditions constraints and stability of power law solutions in $f(R,T)$ gravity. Rao and Neelima (2013a) have discussed perfect fluid Einstein-Rosen Universe in $f(R,T)$ gravity. Rao et al. (2013a) have obtained LRS Bianchi type-I with perfect fluid in a modified theory of gravity and established that the additional condition, special law of variation for the Hubble parameter proposed by Bermann (1983), taken by Adhav (2012) is superfluous. Rao et al. (2014a) have investigated perfect fluid cosmological models in a modified theory of gravity. Recently, Rao and Papa Rao (2015a) have discussed Bianchi type-V dark energy model in $f(R,T)$ gravity. Rao et al. (2015b) have obtained anisotropic Bianchi type-VI_h perfect fluid cosmological models in this theory. Kumar and Singh (2015) have obtained viscous cosmology with matter creation in modified $f(R,T)$ gravity.

Bianchi type cosmological models are important in the sense that these are homogeneous and anisotropic, from which the process of isotropization of the Universe is studied through the passage of time. The simplicity of the field equations and relative ease of solutions made Bianchi space times useful in constructing models of spatially homogeneous and anisotropic cosmologies (Ellis and MacCallum 1969, Ryan and Shepley 1975). The anomalies found in the cosmic microwave background (CMB) and large scale structure observations stimulated a growing interest in anisotropic cosmological models of the Universe. Recently, Reddy et al. (2012a, 2013a), have discussed Bianchi type-*III* cosmological models in $f(R,T)$ theory of gravity. Chaubey and Shukla (2013) have obtained a new class of Bianchi cosmological models in $f(R,T)$ gravity. Mishra and Sahoo (2014) have studied Bianchi type-*VI$_h$* perfect fluid cosmological model in $f(R,T)$ theory. Shamir (2015) has discussed exact solutions of Bianchi type-*V* space-time in $f(R,T)$ gravity. Sahoo and Kumar (2015) have investigated LRS Bianchi type-*I* cosmological model in $f(R,T)$ theory of gravity with $\Lambda(T)$.

Motivated by above discussion and investigations, in this chapter, we have presented spatially homogeneous and anisotropic Bianchi type-*III*, *V* and *VI₀* cosmological models filled with perfect fluid in the framework of $f(R,T)$ gravity proposed by Harko et al. (2011) by considering a specific choice of $f(R,T) = f_1(R) + f_2(T)$.

This chapter is organized as follows: Explicit field equations in $f(R,T)$ gravity are derived with the help of spatially homogeneous and anisotropic Bianchi type-*III*, *V* and *VI₀* space times filled with perfect fluid in section 2.2. Section 2.3 deals with solutions of the field equations. In section 2.4, we have discussed some important features of the models. The last section 2.5 contains some conclusions.

2.2 Metric and Field Equations

We consider a spatially homogeneous Bianchi type-*III*, *V* and *VI₀* metrics of the form

$$ds^2 = dt^2 - A^2 dx^2 - B^2 e^{-2x} dy^2 - C^2 e^{-2mx} dz^2 \qquad (2.2.1)$$

where A, B, C are functions of cosmic time t and m is a constant.

It represents

Bianchi type-*III*   if $m = 0$

Bianchi type-V if $m=1$

Bianchi type-VI_0 if $m=-1$.

The matter tensor for perfect fluid is

$$T_j^i = (\rho + p)u^i u_j - \delta_j^i p \tag{2.2.2}$$

and $u^i = (0,0,0,1)$ is the four velocity in co-moving coordinates which satisfies the condition $u^i u_i = 1$. Here ρ and p are the energy density and pressure of the fluid respectively.

The field equations of $f(R,T)$ gravity for the function $f(R,T) = f_1(R) + f_2(T)$ when the matter source is perfect fluid, as given by Harko et al. (2011), are

$$f_1'(R)R_{ij} - \frac{1}{2}f_1(R)g_{ij} = 8\pi T_{ij} + f_2'(T)T_{ij} + [f_2'(T)p + \tfrac{1}{2}f_2(T)]g_{ij}$$

$$\tag{2.2.3}$$

where $f_1'(R) = \dfrac{\partial f_1(R)}{\partial R}$ & $f_2'(T) = \dfrac{\partial f_2(T)}{\partial T}$.

In this chapter, we consider a particular form of the functions $f_1(R) = \lambda_1 R$ and $f_2(T) = \lambda_2 T$

where λ_1 and λ_2 are any arbitrary parameters, so that

$f(R,T) = \lambda_1 R + \lambda_2 T$.

Then the field equations (2.2.3) will reduce to

$$R_{ij} - \frac{1}{2}Rg_{ij} - \left(p + \frac{T}{2}\right)\frac{\lambda_2}{\lambda_1}g_{ij} = \left(\frac{8\pi + \lambda_2}{\lambda_1}\right)T_{ij} \tag{2.2.4}$$

The field equations (2.2.4) in mixed form can be taken as

$$R_j^i - \frac{1}{2}\delta_j^i R \equiv \left(\frac{8\pi + \lambda_2}{\lambda_1}\right)T_j^i + \frac{\lambda_2}{\lambda_1}\left(p + \frac{T}{2}\right)\delta_j^i \tag{2.2.5}$$

2.3 Solution of the Field Equations

Now with the help of (2.2.2), the field equations (2.2.5) for

the metric (2.2.1) can be written as

$$\frac{\ddot{B}}{B} + \frac{\ddot{C}}{C} + \frac{\dot{B}\dot{C}}{BC} - \frac{m^2}{A^2} = \left(\frac{16\pi + 3\lambda_2}{2\lambda_1}\right)p - \frac{\lambda_2}{2\lambda_1}\rho \tag{2.3.1}$$

$$\frac{\ddot{A}}{A} + \frac{\ddot{C}}{C} + \frac{\dot{A}\dot{C}}{AC} - \frac{m^2}{A^2} = \left(\frac{16\pi + 3\lambda_2}{2\lambda_1}\right)p - \frac{\lambda_2}{2\lambda_1}\rho \tag{2.3.2}$$

$$\frac{\ddot{A}}{A} + \frac{\ddot{B}}{B} + \frac{\dot{A}\dot{B}}{AB} - \frac{1}{A^2} = \left(\frac{16\pi + 3\lambda_2}{2\lambda_1}\right)p - \frac{\lambda_2}{2\lambda_1}\rho \tag{2.3.3}$$

$$\frac{\dot{A}\dot{B}}{AB} + \frac{\dot{B}\dot{C}}{BC} + \frac{\dot{C}\dot{A}}{CA} - \frac{m^2 + m + 1}{A^2} = -\left(\frac{16\pi + 3\lambda_2}{2\lambda_1}\right)\rho + \frac{\lambda_2}{2\lambda_1}p \tag{2.3.4}$$

$$\frac{1}{A^2}\left(\frac{\dot{B}}{B} + m\frac{\dot{C}}{C} - (m+1)\frac{\dot{A}}{A}\right) = 0 \tag{2.3.5}$$

here the overhead dot denotes differentiation with respect to t.

When $m = 0, 1 \, \& -1$ the field equations (2.3.1)-(2.3.5) correspond to the Bianchi type-*III*, *V* and *VI₀* Universes respectively.

2.3.1 Bianchi type-*III* ($m = 0$) model:

If $m = 0$, the field equations (2.3.1) - (2.3.5) will reduce to

$$\frac{\ddot{B}}{B} + \frac{\ddot{C}}{C} + \frac{\dot{B}\dot{C}}{BC} = \left(\frac{16\pi + 3\lambda_2}{2\lambda_1}\right)p - \frac{\lambda_2}{2\lambda_1}\rho \tag{2.3.6}$$

$$\frac{\ddot{A}}{A} + \frac{\ddot{C}}{C} + \frac{\dot{A}\dot{C}}{AC} = \left(\frac{16\pi + 3\lambda_2}{2\lambda_1}\right)p - \frac{\lambda_2}{2\lambda_1}\rho \tag{2.3.7}$$

$$\frac{\ddot{A}}{A} + \frac{\ddot{B}}{B} + \frac{\dot{A}\dot{B}}{AB} - \frac{1}{A^2} = \left(\frac{16\pi + 3\lambda_2}{2\lambda_1}\right)p - \frac{\lambda_2}{2\lambda_1}\rho \tag{2.3.8}$$

$$\frac{\dot{A}\dot{B}}{AB} + \frac{\dot{B}\dot{C}}{BC} + \frac{\dot{C}\dot{A}}{CA} - \frac{1}{A^2} = -\left(\frac{16\pi + 3\lambda_2}{2\lambda_1}\right)\rho + \frac{\lambda_2}{2\lambda_1}p \tag{2.3.9}$$

$$\frac{1}{A^2}\left(\frac{\dot{B}}{B} - \frac{\dot{A}}{A}\right) = 0. \tag{2.3.10}$$

From the equation (2.3.10), we get

$$B = \alpha_1 A.$$

Without loss of generality, by taking $\alpha_1 = 1$, we get

$$B = A. \tag{2.3.11}$$

Using equation (2.3.11), the above field equations (2.3.6) to (2.3.9) will reduce to

$$\frac{\ddot{A}}{A}+\frac{\ddot{C}}{C}+\frac{\dot{A}\dot{C}}{AC}=\left(\frac{16\pi+3\lambda_2}{2\lambda_1}\right)p-\frac{\lambda_2}{2\lambda_1}\rho \qquad (2.3.12)$$

$$2\frac{\ddot{A}}{A}+\frac{\dot{A}^2}{A^2}-\frac{1}{A^2}=\left(\frac{16\pi+3\lambda_2}{2\lambda_1}\right)p-\frac{\lambda_2}{2\lambda_1}\rho \qquad (2.3.13)$$

$$\frac{\dot{A}^2}{A^2}+2\frac{\dot{A}\dot{C}}{AC}-\frac{1}{A^2}=-\left(\frac{16\pi+3\lambda_2}{2\lambda_1}\right)\rho+\frac{\lambda_2}{2\lambda_1}p\; . \qquad (2.3.14)$$

The equations (2.3.12)-(2.3.14) are a system of three independent equations in four unknowns A, C, p and ρ. In order to get a deterministic solution we take the following plausible physical condition, the shear scalar σ is proportional to scalar expansion θ, which leads to the linear relationship between the metric potentials A and C, i.e.

$$C = A^n , \qquad (2.3.15)$$

where $n \neq 0$ is an arbitrary constant.

From equations (2.3.12), (2.3.13) and (2.3.15), we get

$$\frac{\ddot{A}}{A}+(1+n)\frac{\dot{A}^2}{A^2}=\frac{1}{1-n}A^{-2}\; . \qquad (2.3.16)$$

From equation (2.3.16), we get

$$A = \left(k_1 t + k_2\right) \tag{2.3.17}$$

where $k_1^{\,2} = \dfrac{1}{1-n^2}$, $n \neq 1,-1$.

From equations (2.3.15), (2.3.17) and (2.3.11) we have

$$C = \left(k_1 t + k_2\right)^n \tag{2.3.18}$$

and $\quad B = \left(k_1 t + k_2\right)$. $\tag{2.3.19}$

Then the metric (2.2.1) can be written as

$$ds^2 = dt^2 - \left(k_1 t + k_2\right)^2 \left(dx^2 + e^{-2x} dy^2\right) - \left(k_1 t + k_2\right)^{2n} dz^2$$

$$\tag{2.3.20}$$

From equations (2.3.12) to (2.3.14), we get

the pressure as

$$p = \frac{n\lambda_1}{(1-n^2)(k_1 t + k_2)^2}\left(\frac{n+1}{(8\pi + 2\lambda_2)} - \frac{1}{(8\pi + \lambda_2)}\right) \tag{2.3.21}$$

the energy density as

$$\rho = \frac{n\lambda_1}{(n^2 - 1)(k_1 t + k_2)^2}\left(\frac{n+1}{(8\pi + 2\lambda_2)} + \frac{1}{(8\pi + \lambda_2)}\right) \tag{2.3.22}$$

Thus the metric (2.3.20) together with (2.3.21) and (2.3.22) constitutes a Bianchi type-*III* perfect fluid cosmological model in $f(R,T)$ gravity. This is entirely different from the models obtained by Reddy et al. (2012a), Chandel and Ram (2013). Here we have obtained and presented a new class of Bianchi type-*III* perfect fluid cosmological model in $f(R,T)$ gravity.

2.3.2 Bianchi type-*V* ($m = 1$) model:

If $m = 1$, then the equations (2.3.1) - (2.3.5) will reduce to

$$\frac{\ddot{B}}{B} + \frac{\ddot{C}}{C} + \frac{\dot{B}\dot{C}}{BC} - \frac{1}{A^2} = \left(\frac{16\pi + 3\lambda_2}{2\lambda_1}\right)p - \frac{\lambda_2}{2\lambda_1}\rho \qquad (2.3.23)$$

$$\frac{\ddot{A}}{A} + \frac{\ddot{C}}{C} + \frac{\dot{A}\dot{C}}{AC} - \frac{1}{A^2} = \left(\frac{16\pi + 3\lambda_2}{2\lambda_1}\right)p - \frac{\lambda_2}{2\lambda_1}\rho \qquad (2.3.24)$$

$$\frac{\ddot{A}}{A} + \frac{\ddot{B}}{B} + \frac{\dot{A}\dot{B}}{AB} - \frac{1}{A^2} = \left(\frac{16\pi + 3\lambda_2}{2\lambda_1}\right)p - \frac{\lambda_2}{2\lambda_1}\rho \qquad (2.3.25)$$

$$\frac{\dot{A}\dot{B}}{AB} + \frac{\dot{B}\dot{C}}{BC} + \frac{\dot{C}\dot{A}}{CA} - \frac{3}{A^2} = -\left(\frac{16\pi + 3\lambda_2}{2\lambda_1}\right)\rho + \frac{\lambda_2}{2\lambda_1}p \qquad (2.3.26)$$

$$\frac{1}{A^2}\left(\frac{\dot{B}}{B} + \frac{\dot{C}}{C} - \frac{2\dot{A}}{A}\right) = 0. \qquad (2.3.27)$$

From equation (2.3.27), we get

$$A^2 = \alpha_2 \, BC .$$

Without loss of generality, by taking $\alpha_2 = 1$, we get

$$A^2 = BC . \tag{2.3.28}$$

From equations (2.3.24) and (2.3.25), we get

$$\frac{\ddot{B}}{B} - \frac{\ddot{C}}{C} + \frac{\dot{A}}{A}\left(\frac{\dot{B}}{B} - \frac{\dot{C}}{C} \right) = 0 . \tag{2.3.29}$$

In order to get a deterministic solution we take the following plausible physical condition, the shear scalar σ is proportional to scalar expansion θ, which leads to the relationship between metric potentials B and C,

$$\text{i.e., } B = C^n \tag{2.3.30}$$

where $n \neq 0$ is an arbitrary constant.

From equations (2.3.28), (2.3.29) and (2.3.30), we get

$$C = \left(k_5 t + k_6 \right)^{\frac{2}{3(n+1)}} , \quad n \neq -1 \tag{2.3.31}$$

$$B = \left(k_5 t + k_6 \right)^{\frac{2n}{3(n+1)}} \tag{2.3.32}$$

$$A = \left(k_5 t + k_6 \right)^{\frac{1}{3}} \tag{2.3.33}$$

where $k_5 = \dfrac{3}{2}(n+1)k_3$, $k_6 = \dfrac{3}{2}(n+1)k_4$ and k_3 & k_4 are integrating

constants.

Then the metric (2.2.1) can be written as

$$ds^2 = dt^2 - (k_5 t + k_6)^{\frac{2}{3}} dx^2 - e^{-2x}[(k_5 t + k_6)^{\frac{4n}{3(n+1)}} dy^2$$
$$+ (k_5 t + k_6)^{\frac{4}{3(n+1)}} dz^2] .$$

$$(2.3.34)$$

From the field equations (2.3.23) to (2.3.26) we get

the pressure as

$$p = \frac{\lambda_1}{(8\pi + \lambda_2)}\left(\frac{-k_3^2(n^2 + 4n + 1)}{2(k_5 t + k_6)^2} + \frac{1}{(k_5 t + k_6)^{\frac{2}{3}}}\right) - \frac{\lambda_1}{(4\pi + \lambda_2)(k_5 t + k_6)^{\frac{2}{3}}}$$

$$(2.3.35)$$

the energy density as

$$\rho = \frac{\lambda_1}{(8\pi + \lambda_2)}\left(\frac{-k_3^2(n^2 + 4n + 1)}{2(k_5 t + k_6)^2} + \frac{1}{(k_5 t + k_6)^{\frac{2}{3}}}\right) + \frac{\lambda_1}{(4\pi + \lambda_2)(k_5 t + k_6)^{\frac{2}{3}}}$$

$$(2.3.36)$$

Thus the metric (2.3.34) together with (2.3.35) and (2.3.36) constitutes a Bianchi type-V perfect fluid cosmological model in case of $f(R,T)$ gravity. This is entirely different from the models obtained by Ahmed and Pradhan (2014).

2.3.3 Bianchi type-VI_0 ($m = -1$) model:

If $m = -1$, then the field equations (2.3.1) - (2.3.5) will reduce to

$$\frac{\ddot{B}}{B} + \frac{\ddot{C}}{C} + \frac{\dot{B}\dot{C}}{BC} + \frac{1}{A^2} = \left(\frac{16\pi + 3\lambda_2}{2\lambda_1}\right)p - \frac{\lambda_2}{2\lambda_1}\rho \qquad (2.3.37)$$

$$\frac{\ddot{A}}{A} + \frac{\ddot{C}}{C} + \frac{\dot{A}\dot{C}}{AC} - \frac{1}{A^2} = \left(\frac{16\pi + 3\lambda_2}{2\lambda_1}\right)p - \frac{\lambda_2}{2\lambda_1}\rho \qquad (2.3.38)$$

$$\frac{\ddot{A}}{A} + \frac{\ddot{B}}{B} + \frac{\dot{A}\dot{B}}{AB} - \frac{1}{A^2} = \left(\frac{16\pi + 3\lambda_2}{2\lambda_1}\right)p - \frac{\lambda_2}{2\lambda_1}\rho \qquad (2.3.39)$$

$$\frac{\dot{A}\dot{B}}{AB} + \frac{\dot{B}\dot{C}}{BC} + \frac{\dot{C}\dot{A}}{CA} - \frac{1}{A^2} = -\left(\frac{16\pi + 3\lambda_2}{2\lambda_1}\right)\rho + \frac{\lambda_2}{2\lambda_1}p \qquad (2.3.40)$$

$$\frac{1}{A^2}\left(\frac{\dot{B}}{B} - \frac{\dot{C}}{C}\right) = 0. \qquad (2.3.41)$$

From equation (2.3.41), we get

$$B = \alpha_3 C.$$

Without loss of generality, by taking $\alpha_3 = 1$, we get

$$B = C. \qquad (2.3.42)$$

Using (2.3.42), the above field equations (2.3.37) to (2.3.40) will

reduce to

$$\frac{2\ddot{C}}{C} + \frac{\dot{C}^2}{C^2} + \frac{1}{A^2} = \left(\frac{16\pi + 3\lambda_2}{2\lambda_1}\right)p - \frac{\lambda_2}{2\lambda_1}\rho \qquad (2.3.43)$$

$$\frac{\ddot{A}}{A}+\frac{\ddot{C}}{C}+\frac{\dot{A}\dot{C}}{AC}-\frac{1}{A^2}=\left(\frac{16\pi+3\lambda_2}{2\lambda_1}\right)p-\frac{\lambda_2}{2\lambda_1}\rho \qquad (2.3.44)$$

$$\frac{\dot{A}\dot{C}}{AC}+\frac{\dot{C}^2}{C^2}+\frac{\dot{A}\dot{C}}{AC}-\frac{1}{A^2}=-\left(\frac{16\pi+3\lambda_2}{2\lambda_1}\right)\rho+\frac{\lambda_2}{2\lambda_1}p. \qquad (2.3.45)$$

The equations (2.3.43)-(2.3.45) is a system of three equations in four unknowns $A,C,\ p$ and ρ. In order to get a deterministic solution we take the following plausible physical condition, the shear scalar σ is proportional to scalar expansion θ, which leads to the linear relationship between the metric potentials A and C, i.e.

$$A = C^n \qquad (2.3.46)$$

where $n \neq 0$ is an arbitrary constant.

From the equations (2.3.43), (2.3.44) and (2.3.46), we get

$$\ddot{C}+(n+1)\frac{\dot{C}^2}{C}=\frac{2}{(n-1)}\left(\frac{1}{C^{2n-1}}\right). \qquad (2.3.47)$$

From equation (2.3.47), we get

$$C = \left[n(k_7 t + k_8)\right]^{\frac{1}{n}} \qquad (2.3.48)$$

where $k_7{}^2 = \dfrac{1}{n-1}$, $n \neq 1$ and k_8 is an integrating constant.

From equations (2.3.42), (2.3.46) and (2.3.48) we have

$$B = \left[n(k_7 t + k_8)\right]^{\frac{1}{n}} \tag{2.3.49}$$

$$A = n(k_7 t + k_8). \tag{2.3.50}$$

Then the metric (2.2.1) can be written as

$$ds^2 = dt^2 - \left[n(k_7 t + k_8)\right]^2 dx^2 - \left[n(k_7 t + k_8)\right]^{\frac{2}{n}}\left(e^{-2x} dy^2 + e^{2x} dz^2\right)$$

$$\tag{2.3.51}$$

From the field equations (2.3.43) to (2.3.45), we get

the pressure as

$$p = \frac{\lambda_1}{n^2(n-1)(k_7 t + k_8)^2}\left(\frac{1}{4\pi + \lambda_2} - \frac{n}{8\pi + \lambda_2}\right) \tag{2.3.52}$$

the energy density as

$$\rho = \frac{\lambda_1}{n^2(1-n)(k_7 t + k_8)^2}\left(\frac{1}{4\pi + \lambda_2} + \frac{n}{8\pi + \lambda_2}\right). \tag{2.3.53}$$

Thus the metric (2.3.51) together with (2.3.52) and (2.3.53) constitutes a Bianchi type-VI_0 perfect fluid cosmological model in case of $f(R,T)$ gravity. For particular values of $\lambda_1 = 1$ and $\lambda_2 = 2\lambda$, we will get back to the model obtained by Rao and Neelima (2013b). Ramifications

2.4 Some Important Features of the Models

2.4.1 Bianchi type-*III* ($m = 0$) model:

The volume element of the model (2.3.20) is given by

$$V = (-g)^{\frac{1}{2}} = (k_1 t + k_2)^{n+2} e^{-x} .$$
(2.4.1)

Average scale factor a is given by

$$a = V^{\frac{1}{3}} = (k_1 t + k_2)^{\frac{n+2}{3}} e^{\frac{-x}{3}}$$
(2.4.2)

The expansion scalar θ is given by

$$\theta = u^i_{,i} = \frac{(n+2)k_1}{k_1 t + k_2} .$$
(2.4.3)

The shear scalar σ^2 is given by

$$\sigma^2 = \frac{7}{18}\left(\frac{(n+2)k_1}{k_1 t + k_2}\right)^2$$
(2.4.4)

Mean Hubble parameter H is given by

$$H = \frac{1}{3}(2H_1 + H_2) = \left(\frac{n+2}{k_1 t + k_2}\right)\frac{k_1}{3}$$
(2.4.5)

The deceleration parameter q is given by

$$q = -3\theta^{-2}(\theta_{,i} u^i + \frac{1}{3}\theta^2) = \frac{1-n}{n+2} , \quad n \neq -2.$$
(2.4.6)

The mean anisotropy parameter A_m is given by

$$A_m = \frac{1}{3}\sum_{i=1}^{3}\left(\frac{H_i - H}{H}\right)^2 = \frac{2(n-1)^2}{(n+2)} \tag{2.4.7}$$

The overall density parameter Ω is given by

$$\Omega = \frac{\rho}{3H^2} = \frac{-3n\lambda_1}{(n+2)^2}\left(\frac{n+1}{8\pi + 2\lambda_2} + \frac{1}{8\pi + \lambda_2}\right) \tag{2.4.8}$$

Red shift Z is given by

$$Z = \frac{1}{a(t)} - 1 = \left(k_1 t + k_2\right)^{-\left(\frac{n+2}{3}\right)} e^{\frac{x}{3}} - 1 \tag{2.4.9}$$

Jerk parameter is given by

$$j = \frac{1}{H^3}\frac{\ddot{a}}{a} = \frac{(n-1)(n-4)}{(n+2)^2} \tag{2.4.10}$$

Luminosity parameter $d_L = r_1 a_0 (1+z)$, where $r_1 = \int^{t_0} \frac{1}{a(t)} dt$

$$d_L = \frac{3}{(1-n)k_1} e^{\frac{2x}{3}}\left[\left(k_1 t_0 + k_2\right)^{\frac{1-n}{3}}\left(k_1 t + k_2\right)^{\frac{-(n+2)}{3}} - \left(k_1 t + k_2\right)^{\frac{-(1+2n)}{3}}\right]$$

$$\tag{2.4.11}$$

The distance modulus $D(Z) = 5\log d_L + 25$

$$D(Z) = 5\log\left\{\frac{3}{(1-n)k_1} e^{\frac{2x}{3}}\left[\left(k_1 t_0 + k_2\right)^{\frac{1-n}{3}}\left(k_1 t + k_2\right)^{\frac{-(n+2)}{3}} - \left(k_1 t + k_2\right)^{\frac{-(1+2n)}{3}}\right]\right\} + 25$$

$$\tag{2.4.12}$$

2.4.2 Bianchi type-V ($m=1$) model:

The spatial volume for the model (2.3.34) is

$$V = \sqrt{-g} = (k_5 t + k_6)e^{-2x} \tag{2.4.13}$$

Average scale factor a is given by

$$a(t) = V^{\frac{1}{3}} = (k_5 t + k_6)^{\frac{1}{3}} e^{\frac{-2x}{3}} \tag{2.4.14}$$

The expression for expansion scalar θ calculated for the flow

vector u^i is given by

$$\theta = u^i{}_{,i} = \frac{k_5}{k_5 t + k_6} \tag{2.4.15}$$

and the shear σ^2 is given by

$$\sigma^2 = \frac{1}{2}\sigma^{ij}\sigma_{ij} = \frac{7}{18}\left(\frac{k_5}{k_5 t + k_6}\right)^2 \tag{2.4.16}$$

The deceleration parameter q is given by

$$q = (-3\theta^{-2})(\theta_{,i}u^i + \frac{1}{3}\theta^2) = 2 \tag{2.4.17}$$

The Hubble's parameter H is given by

$$H = \frac{k_5}{3(k_5 t + k_6)} \tag{2.4.18}$$

The mean anisotropy parameter A_m is given by

$$A_m = \frac{1}{3}\sum_{i=1}^{3}\left(\frac{H_i - H}{H}\right)^2 = \frac{2}{3}(n-1)^2 \tag{2.4.19}$$

The overall density parameter Ω is given by

$$\Omega = \frac{\rho}{3H^2} = \frac{\lambda_1}{(8\pi + \lambda_2)}\left(\frac{-2(n^2 + 4n + 1)}{3(n+1)^2} + \frac{3}{k_5^2}(k_5 t + k_6)^{4/3}\right)$$
$$+ \frac{3\lambda_1}{k_5^2(4\pi + \lambda_2)}(k_5 t + k_6)^{4/3} \tag{2.4.20}$$

Red shift Z is given by

$$Z = \frac{1}{a(t)} - 1 = (k_5 t + k_6)^{\frac{-1}{3}} e^{\frac{2x}{3}} - 1 \tag{2.4.21}$$

Jerk parameter is given by

$$j = \frac{1}{H^3}\left(\frac{\ddot{a}}{a}\right) = 10 \tag{2.4.22}$$

Luminosity parameter $d_L = r_1 a_0 (1 + z)$, where $r_1 = \int^{t_0} \frac{1}{a(t)} dt$

$$d_L = \frac{3}{2k_5^2}\left[(k_5 t_0 + k_6)^{\frac{2}{3}}(k_5 t + k_6)^{\frac{-1}{3}} - (k_5 t + k_6)^{\frac{1}{3}}\right] \tag{2.4.23}$$

The distance modulus $D(Z) = 5\log d_L + 25$

$$D(Z) = 5\log\left\{\frac{3}{2k_5^2}\left[(k_5 t_0 + k_6)^{\frac{2}{3}}(k_5 t + k_6)^{\frac{-1}{3}} - (k_5 t + k_6)^{\frac{1}{3}}\right]\right\} + 25 \tag{2.4.24}$$

2.4.3 Bianchi type-VI_0 ($m = -1$) model:

The spatial volume for the model (2.3.51) is

$$V = (-g)^{1/2} = \left(n(k_7 t + k_8)\right)^{\frac{n+2}{n}}$$

(2.4.25)

Average scale factor a is given by

$$a(t) = V^{1/3} = \left[n(k_7 t + k_8)\right]^{\frac{n+2}{3n}}$$

(2.4.26)

The expression for expansion scalar θ calculated for the flow

vector u^i is given by

$$\theta = u^i{}_{,i} = \frac{k_7(n+2)}{n(k_7 t + k_8)}$$

(2.4.27)

and the shear σ^2 is given by

$$\sigma^2 = \frac{1}{2}\sigma^{ij}\sigma_{ij} = \frac{7}{18}\left(\frac{k_7(n+2)}{n(k_7 t + k_8)}\right)^2$$

(2.4.28)

The deceleration parameter q is given by

$$q = (-3\theta^{-2})(\theta_{,i}u^i + \frac{1}{3}\theta^2) = \frac{2(n-1)}{n+2}$$

(2.4.29)

The Hubble's parameter H is given by

$$H = \frac{k_7(n+2)}{3n(k_7 t + k_8)}$$

(2.4.30)

The overall density parameter Ω is given by

$$\Omega = \frac{\rho}{3H^2} = \frac{-3\lambda_1}{(n+2)^2}\left(\frac{1}{4\pi+\lambda_2}+\frac{n}{8\pi+\lambda_2}\right)$$

(2.4.31)

The mean anisotropy parameter A_m is given by

$$A_m = \frac{1}{3}\sum_{i=1}^{3}\left(\frac{H_i-H}{H}\right)^2 = 2\left(\frac{n-1}{n+2}\right)^2$$

(2.4.32)

where $\Delta H_i = H_i - H \quad (i=1,2,3)$

Red Shift Z is given by

$$Z = \frac{1}{a(t)}-1 = \left[n(k_7 t + k_8)\right]^{-\left(\frac{n+2}{3n}\right)}-1$$

(2.4.33)

Jerk parameter is given by

$$j = \frac{1}{H^3}\left(\frac{\ddot{a}}{a}\right) = \frac{2(n-1)(5n-2)}{(n+2)^2}$$

(2.4.34)

Luminosity parameter $d_L = r_1 a_0(1+z)$, where $r_1 = \int^{t_0}\frac{1}{a(t)}dt$

$$d_L = \frac{3n^{\frac{n-4}{3n}}}{2k_7(n-1)}\left[(k_7 t_0 + k_8)^{\frac{-(n+2)}{3n}}(k_7 t + k_8)^{\frac{2(n-1)}{3n}}-(k_7 t + k_8)^{\frac{n-4}{3n}}\right]$$

(2.4.35)

The distance modulus $D(Z) = 5\log d_L + 25$

$$D(Z) = 5\log\left\{\frac{3n^{\frac{n-4}{3n}}}{2k_7(n-1)}\left[(k_7t_0+k_8)^{\frac{-(n+2)}{3n}}(k_7t+k_8)^{\frac{2(n-1)}{3n}} - (k_7t+k_8)^{\frac{n-4}{3n}}\right]\right\} + 25$$

$$(2.4.36)$$

2.5 Conclusions

In this chapter, we have presented spatially homogeneous and anisotropic Bianchi type-*III*, *V* and *VI₀* cosmological models filled with perfect fluid in the framework of $f(R,T)$ gravity proposed by Harko et al. (2011) with an appropriate choice of a function $f(R,T) = f_1(R) + f_2(T)$.

The following are the observations and conclusions:

	Bianchi type-*III*	Bianchi type-*V*	Bianchi type-*VI₀*
Singularity	No singularity for $n > 0$	No singularity for $n > 0$	No singularity for $n > 0$
Spatial volume (V)	at $t = \frac{-k_2}{k_1}$, the spatial volume vanishes and increases	at $t = \frac{-k_6}{k_5}$, the spatial volume vanishes	at $t = \frac{-k_8}{k_7}$, the spatial volume vanishes and increases

	continuously with time for $n > -2$	and increases continuously with time	continuously with time
Expansion scalar θ	Decreases with the increase of time and diverges at $t = \frac{-k_2}{k_1}$	Decreases with the increase of time and diverges at $t = \frac{-k_6}{k_5}$	Decreases with the increase of time and diverges at $t = \frac{-k_8}{k_7}$
Shear scalar σ^2	Decreases with the increase of time and diverges at $t = \frac{-k_2}{k_1}$	Decreases with the increase of time and diverges at $t = \frac{-k_6}{k_5}$	Decreases with the increase of time and diverges at $t = \frac{-k_8}{k_7}$
Hubble parameter H	Decreases with the increase of time and diverges at $t = \frac{-k_2}{k_1}$	Decreases with the increase of time and diverges at $t = \frac{-k_6}{k_5}$	Decreases with the increase of time and diverges at $t = \frac{-k_8}{k_7}$

Energy density ρ, pressure p	Tends to zero as $t \to \infty$ and diverges at $t = \frac{-k_2}{k_1}$	Tends to zero as $t \to \infty$ and diverges at $t = \frac{-k_6}{k_5}$	Tends to zero as $t \to \infty$ and diverges at $t = \frac{-k_8}{k_7}$
Deceleration parameter q	$q < 0$ except for $-2 < n < 1$, it represents accelerating Universe	$q > 0$, it represents decelerating Universe	$q < 0$ for $-2 < n < 1$, it represents accelerating Universe
Mean anisotropy parameter A_m	$A_m \neq 0$, indicates that this model is always anisotropic	$A_m = 0$, for $n = 1$ indicates that this model is isotropic	$A_m \neq 0$, indicates that this model is always anisotropic
Jerk Parameter j	$j = \frac{(n-1)(n-4)}{(n+2)^2}$	$j = 10$	$j = \frac{2(n-1)(5n-2)}{(n+2)^2}$

Since $q > 0$, for the Bianchi type-V model, it decelerates in the standard way. However, in spite of the fact that the Universe, in this case, decelerates in the standard way it will accelerate in finite time due to cosmic re collapse where the Universe in turns inflates "decelerates and then accelerates" (Nojiri and Odintsov 2003b). However, for Bianchi type-*III* and *VI₀* models the deceleration parameter q is negative, so they represent accelerating expansion of the Universes.

Chapter-3

Bianchi type-*II*, *VIII* and *IX* perfect fluid Universes in $f(R,T)$ gravity*

*The work presented in this chapter is communicated to following journal:

➤ Perfect fluid Bianchi Universes in a modified theory of gravity: Proceedings of the National Academy of Sciences journal

3.1 Introduction

The aim of modern cosmology is to determine the large-scale structure of the Universe. The astronomical observations of type-I a supernovae experiments Reiss et al. (1998), Perlmutter et al. (1999), Bennet et al. (2003) suggest that the observable Universe is undergoing an accelerated expansion. Also, observations such as cosmic microwave background radiation (Spergel et al. 2007) and large-scale structure (Tegmark et al. 2004b) provide an indirect evidence for the late time accelerated expansion of the Universe. It is generally believed that some sort of 'dark energy' is pervading the whole Universe. This is a hypothetical form of energy that permeates all of the space and tends to increase the rate of expansion of the Universe (Peebles and Ratra 2003). The dark energy is a prime candidate for explaining the recent cosmic observations. In recent years, there has been a lot of interest in alternative theories of gravitation. In view of the late time acceleration of the Universe and the existence of the dark matter and dark energy, recently, modified theories of gravity have been developed. Noteworthy amongst them are $f(R)$ theory of gravity formulated by Nojiri and Odintsov (2003a) and $f(R,T)$ gravity proposed by Harko et al. (2011). Bertolami et al. (2007) proposed a generalization of $f(R)$ theory of gravity by including in the theory an explicit coupling of an arbitrary function of the Ricci scalar R with the matter Lagrangian density L_m. Harko et al. developed this modified theory of gravity, where the gravitational Lagrangian is given by an arbitrary function of the Ricci scalar R and of the trace T of the stress-energy tensor. They have obtained the gravitational field equations in the metric formalism, as well as, the equations of motion for test particles, which follow from the covariant divergence of the stress-energy tensor. Nojiri et al. (2007) developed a general programme for the unification of matter-dominated era with acceleration epoch for scalar-tensor theory or dark fluid. Shamir (2010) proposed a physically viable $f(R)$ gravity model, which showed the unification of early time inflation and late time acceleration.

Sharif and Zubair (2012b) have investigated thermodynamics in $f(R,T)$ theory of gravity. Sharif and Zubair (2013b) have investigated cosmology of holographic and new age graphic $f(R,T)$ models. Rao and Neelima (2013b) have discussed perfect fluid Bianchi type-VI_0 Universe in $f(R,T)$ theory of gravitation. Rao et al. (2013b) have discussed perfect fluid cosmological models in general relativity and $f(R,T)$ gravity, which was again discussed by Sahoo et al. (2014). Chakroborthy (2013) has discussed $f(R,T)$ gravity by considering three different cases. Rao and Suryanarayana (2014) have discussed higher dimensional perfect fluid cosmological models in this theory. Recently, Rao et al. (2015c) have obtained Bianchi type-III, V and VI_0 Universes in this theory. Rao and Suryanarayana (2015) have investigated Kantowski-Sachs cosmological model in this theory.

The study of familiar solutions like FRW Universe with positive curvature, the de Sitter Universe, the Taub- Nut solutions etc. correspond to Bianchi type-*II*, *VIII* and *IX* space-times. Rao et al. (2008a, b, c) have studied Bianchi type-*II*, *VIII* and *IX* cosmological models in different theories of gravitation. Rao and Vijaya Santhi (2012a) have obtained Bianchi type-*II*, *VIII* and *IX* magnetized cosmological models in Brans-Dicke theory of gravitation. Rao and Sireesha (2012a), Rao et al. (2012a) have studied Bianchi type-*II*, *VIII* and *IX* string cosmological models with bulk viscosity in some scalar-tensor theories of gravitation. Recently, Singh and Sharma (2014a) have discussed Bianchi type-*II* dark energy model in $f(R,T)$ gravity. Rao et al. (2014a) have discussed Bianchi type-*II*, *VIII* and *IX* cosmological models in $f(R,T)$ theory of gravity.

Inspired by the above investigations and discussion, in this chapter, we focus our attention on investigating spatially homogeneous and anisotropic Bianchi type-*II*, *VIII* and *IX* cosmological models filled with perfect fluid in the framework of $f(R,T)$ gravity with a general choice of the function $f(R,T) = f_1(R) + f_2(T)$.

The present chapter is organized as follows: In section 3.2, we have obtained the explicit field equations in the $f(R,T)$ theory of gravity with the help of Bianchi type-*II*, *VIII* and *IX* metric in presence of perfect fluid. In section 3.3, we have obtained solutions of the field equations. Some important features are presented in section 3.4. The last section 3.5 contains some conclusions.

3.2 Metric and Field Equations

We consider spatially homogeneous Bianchi type-*II*, *VIII* and *IX* metrics in the form

$$ds^2 = dt^2 - R^2\left[d\theta^2 + f^2(\theta)d\varphi^2\right] - S^2[d\psi + h(\theta)d\varphi]^2 \qquad (3.2.1)$$

where θ, φ and ψ are the Eulerian angles. Also R and S are functions of t only.

It represents

Bianchi type-*II*   if $f(\theta)=1$ and $h(\theta)=\theta$

Bianchi type-*VIII* if $f(\theta)=Cosh\,\theta$ and $h(\theta)=Sinh\,\theta$

Bianchi type-*IX*   if $f(\theta)=Sin\,\theta$ and $h(\theta)=Cos\,\theta$

The matter tensor for perfect fluid is

$$T_j^i = (\rho + p)u^i u_j - \delta_j^i p \qquad\qquad (3.2.2)$$

and $u^i = (0,0,0,1)$ is the four velocity in co-moving coordinates which satisfies the condition $u^i u_i = 1$. Here ρ and p are the energy density and pressure of the fluid respectively.

In this chapter, we are considering the following special case given by Harko et al. (2011)

$$f(R,T) = f_1(R) + f_2(T). \qquad\qquad (3.2.3)$$

For this particular form of $f(R,T)$, the gravitational field equations as given by Harko et al. (2011) are

$$f_1'(R)R_{ij} - \frac{1}{2}f_1(R)g_{ij} = 8\pi T_{ij} + f_2'(T)T_{ij} + [f_2'(T)p + \tfrac{1}{2}f_2(T)]g_{ij} \qquad (3.2.4)$$

where the prime indicates derivative with respect to the argument.

Here we consider

$$f_1(R) = \lambda_1 R \text{ and } f_2(T) = \lambda_2 T, \qquad\qquad (3.2.5)$$

where λ_1 and λ_2 are an arbitrary constants.

Now the field equations (3.2.4) due to (3.2.5) become

$$G_j^i \equiv R_j^i - \frac{1}{2}\delta_j^i R \equiv \left(\frac{8\pi + \lambda_2}{\lambda_1}\right)T_j^i + \frac{\lambda_2}{\lambda_1}\left(p + \frac{T}{2}\right)\delta_j^i \qquad (3.2.6)$$

3.3 Solution of Field Equations

Now with the help of (3.2.2) the field equations (3.2.5) for the metric (3.2.1) can be written as

$$\frac{\ddot{R}}{R} + \frac{\ddot{S}}{S} + \frac{\dot{R}\dot{S}}{RS} + \frac{S^2}{4R^4} = \left(\frac{8\pi + \lambda_2}{\lambda_1}\right)p - \left(\frac{\rho - p}{2}\right)\frac{\lambda_2}{\lambda_1} \qquad (3.3.1)$$

$$2\frac{\ddot{R}}{R} + \frac{\dot{R}^2 + \delta}{R^2} - \frac{3S^2}{4R^4} = \left(\frac{8\pi + \lambda_2}{\lambda_1}\right)p - \left(\frac{\rho - p}{2}\right)\frac{\lambda_2}{\lambda_1} \qquad (3.3.2)$$

$$2\frac{\dot{R}\dot{S}}{RS} + \frac{\dot{R}^2 + \delta}{R^2} - \frac{S^2}{4R^4} = -\left(\frac{8\pi + \lambda_2}{\lambda_1}\right)\rho - \left(\frac{\rho - p}{2}\right)\frac{\lambda_2}{\lambda_1} \qquad (3.3.3)$$

here the over head dot denotes differentiation with respect to t.

The field equations (3.3.1) to (3.3.3) are only three independent equations with four unknowns R, S, ρ and p. So, in order to get a deterministic solution we take the following plausible physical condition, the shear scalar σ is proportional to scalar expansion θ, which leads to the linear relationship between the metric potentials R and S, i.e.,

$$S = R^n \tag{3.3.4}$$

where n is an arbitrary constant.

From equations (3.3.1), (3.3.2) and (3.3.4), we get

$$\frac{\ddot{R}}{R} + (1+n)\frac{\dot{R}^2}{R^2} + \frac{\delta}{(1-n)R^2} - \frac{R^{2n-4}}{(1-n)} = 0 \ , \ n \neq 1 \tag{3.3.5}$$

3.3.1 Bianchi type-II ($\delta = 0$) model:

If $\delta = 0$, from equation (3.3.5) we get

$$\frac{\ddot{R}}{R} + (1+n)\frac{\dot{R}^2}{R^2} - \frac{R^{2n-4}}{(1-n)} = 0 \ , \ n \neq 1 \tag{3.3.6}$$

From equation (3.3.6), we get

$$R = (at+b)^{1/2-n} \ , \ n \neq 2 \tag{3.3.7}$$

where a and n are related by $2a^2n(1-n)-(2-n)^2 = 0$

and $a \neq 0, b$ are arbitrary constants.

From equations (3.3.4) and (3.3.7), we get

$$S = (at+b)^{n/2-n} \tag{3.3.8}$$

From equations (3.3.1)-(3.3.8), we get

the energy density as

$$\rho = \frac{\lambda_1}{2(2-n)^2(at+b)^2}\left[\frac{a^2(n+1)(n-2)}{(8\pi+\lambda_2)} - \frac{[2na^2(n+3)-1]}{(16\pi+4\lambda_2)}\right] \qquad (3.3.9)$$

and the pressure as

$$p = \frac{\lambda_1}{2(2-n)^2(at+b)^2}\left[\frac{2na^2(n+3)-1}{(16\pi+4\lambda_2)} + \frac{a^2(n+1)(n-2)}{(8\pi+\lambda_2)}\right] \qquad (3.3.10)$$

The metric (3.2.1), in this case, can be written as

$$ds^2 = dt^2 - (at+b)^{\frac{2}{2-n}}\left[d\theta^2 + d\varphi^2\right] - (at+b)^{\frac{2n}{2-n}}[d\psi + \theta d\varphi]^2 \qquad (3.3.11)$$

Thus the metric (3.3.11) together with (3.3.9) and (3.3.10)

constitutes a spatially homogeneous and anisotropic Bianchi type-

II perfect fluid cosmological model in $f(R,T)$ gravity.

3.3.2 Bianchi type-*VIII* ($\delta = -1$) model:

If $\delta = -1$, from equation (3.3.5) we get

$$\frac{\ddot{R}}{R} + (1+n)\frac{\dot{R}^2}{R^2} - \frac{1}{(1-n)R^2} - \frac{R^{2n-4}}{(1-n)} = 0 \ , \qquad n \neq 1 \qquad (3.3.12)$$

we can solve the above equation and get the deterministic solution
only for $n = 2$.

With a suitable substitution, the equation (3.3.12) can be written
as

$$\dot{R}^2 = w^2 - \gamma^2 R^2 \tag{3.3.13}$$

where $w^2 = \dfrac{-1}{3}$ & $\gamma^2 = \dfrac{1}{4}$

From equation (3.3.13), we get

$$R^2 = \dfrac{-4}{3} Sin^2 \dfrac{t}{2} \tag{3.3.14}$$

From equations (3.3.4) and (3.3.14), we get

$$S^2 = \dfrac{16}{9} Sin^4 \dfrac{t}{2} \tag{3.3.15}$$

From equations (3.3.1), (3.3.3), (3.3.14) and (3.3.15), we get

the energy density as

$$\rho = \dfrac{-\lambda_1 Co\sec^2 \frac{t}{2}}{2(8\pi + \lambda_2)} - \dfrac{3\lambda_1 Cot^2 \frac{t}{2}}{4(4\pi + \lambda_2)} \tag{3.3.16}$$

and the pressure as

$$p = \dfrac{3\lambda_1 Cot^2 \frac{t}{2}}{4(4\pi + \lambda_2)} - \dfrac{\lambda_1 Co\sec^2 \frac{t}{2}}{2(8\pi + \lambda_2)} \tag{3.3.17}$$

The metric (3.2.1), in this case can be written as

$$ds^2 = dt^2 + \tfrac{4}{3} Sin^2 \tfrac{t}{2}\left(d\theta^2 + Cosh^2\theta\, d\varphi^2\right) - \tfrac{16}{9} Sin^4 \tfrac{t}{2}\left(d\psi + Sinh\theta\, d\varphi\right)^2$$

(3.3.18)

Thus the metric (3.3.18) together with (3.3.16) and (3.3.17) constitutes a spatially homogeneous and anisotropic Bianchi type-*VIII* perfect fluid cosmological model in $f(R,T)$ gravity.

3.3.3 Bianchi type-*IX* ($\delta=1$) model:

If $\delta=1$, from equation (3.3.5) we get

$$\frac{\ddot{R}}{R} + (1+n)\frac{\dot{R}^2}{R^2} + \frac{1}{(1-n)R^2} - \frac{R^{2n-4}}{(1-n)} = 0, \quad n \neq 1.$$

(3.3.19)

We can solve the above equation and get the deterministic solution only for $n=2$. With a suitable substitution, the equation (3.3.19) can be written as

$$\dot{R}^2 = w^2 - \gamma^2 R^2$$

(3.3.20)

where $w^2 = \tfrac{1}{3}$ & $\gamma^2 = \tfrac{1}{4}$

From equation (3.3.20), we get

$$R^2 = \tfrac{4}{3} Sin^2 \tfrac{t}{2}$$

(3.3.21)

From equations (3.3.4) and (3.3.21), we get

$$S^2 = \tfrac{16}{9} Sin^4 \tfrac{t}{2}$$

$$(3.3.22)$$

From equations (3.3.1), (3.3.3), (3.3.21) and (3.3.22), we get

the energy density

$$\rho = \frac{-3\lambda_1 Cot^2 \dfrac{t}{2}}{4(4\pi + \lambda_2)} - \frac{\lambda_1 Co\sec^2 \dfrac{t}{2}}{2(8\pi + \lambda_2)}$$

$$(3.3.23)$$

and the pressure

$$p = \frac{3\lambda_1 Cot^2 \dfrac{t}{2}}{4(4\pi + \lambda_2)} - \frac{\lambda_1 Co\sec^2 \dfrac{t}{2}}{2(8\pi + \lambda_2)}$$

$$(3.3.24)$$

The metric (3.2.1), in this case can be written as

$$ds^2 = dt^2 - \tfrac{4}{3} Sin^2 \tfrac{t}{2}\left(d\theta^2 + Sin^2\theta \, d\varphi^2\right) - \tfrac{16}{9} Sin^4 \tfrac{t}{2}(d\psi + Cos\theta \, d\varphi)^2$$

$$(3.3.25)$$

Thus the metric (3.3.25) together with (3.3.23) and (3.3.24) constitutes a spatially homogeneous and anisotropic Bianchi type-IX perfect fluid cosmological model in $f(R,T)$ gravity.

3.4 Some Important Features of the Models

3.4.1 Bianchi type-*II* ($\delta = 0$) model:

The spatial volume for the model (3.3.11) is

$$V = (at+b)^{\frac{n+2}{2-n}}$$

(3.4.1)

The average scale factor

$$a(t) = V^{\frac{1}{3}} = (at+b)^{\frac{n+2}{3(2-n)}}$$

(3.4.2)

The expression for expansion scalar θ calculated for the flow vector u^i is given by

$$\theta = 3H = \left(\frac{n+2}{2-n}\right)\frac{a}{(at+b)}$$

(3.4.3)

and the shear σ is given by

$$\sigma^2 = \frac{7}{18}\left(\frac{n+2}{2-n}\right)^2 \frac{a^2}{(at+b)^2}$$

(3.4.4)

The deceleration parameter q is given by

$$q = \frac{d}{dt}\left(\frac{1}{H}\right) - 1 = \frac{-4(n-1)}{(n+2)}, n \neq 1, 2 \,\&\, -2$$

(3.4.5)

The recent observations of SN Ia reveal that the present Universe is accelerating and the value of deceleration parameter lies somewhere in the range $-1 < q < 0$. From equation (3.4.5) we can see that for $1 < n < 2$, q is in the above range which may be attributed to the current accelerated power law expansion of the Universe.

Also $q < -1$, for $n > 2$ and hence it represents super exponential expansion of the Universe.

The components of directional Hubble parameters H_x, H_y & H_z are given by

$$H_x = H_y = \frac{\dot{R}}{R} = \left(\frac{1}{2-n}\right)\frac{a}{(at+b)}, \quad H_z = \frac{\dot{S}}{S} = \left(\frac{n}{2-n}\right)\frac{a}{(at+b)}$$

Therefore the generalized mean Hubble parameter (H) is

$$H = \frac{1}{3}\left(H_x + H_y + H_z\right) = \frac{1}{3}\left(\frac{n+2}{2-n}\right)\frac{a}{(at+b)} \tag{3.4.6}$$

The average anisotropy parameter is defined by

$$A_m = \frac{1}{3}\sum_{i=1}^{3}\left(\frac{H_i - H}{H}\right)^2 = \frac{2(n-1)^2}{(n+2)^2}, \quad \text{where } i = 1,2,3 \tag{3.4.7}$$

Red shift

$$Z = \frac{1}{a(t)} - 1 = (at + b)^{\frac{n+2}{3(n-2)}} - 1 \tag{3.4.8}$$

The jerk parameter is

$$j = \frac{1}{H^3}\frac{\ddot{a}}{a} = \frac{4(n-1)(7n-10)}{(n+2)^2} \tag{3.4.9}$$

Luminosity parameter $d_L = r_1(1+z)$, where $r_1 = \int\limits_t^{t_0} \frac{1}{a(t)}\, dt$

$$d_L = \frac{3(2-n)}{4(1-n)a}\left[(at_0 + b)^{\frac{4(1-n)}{3(2-n)}}(at + b)^{\frac{-(n+2)}{3(2-n)}} - (at + b)^{\frac{2-5n}{3(2-n)}}\right]$$

$$\tag{3.4.10}$$

3.4.2 Bianchi type-*VIII* ($\delta = -1$) and *IX* ($\delta = 1$) models:

The spatial volume for both the models (3.3.18) and (3.3.25) is

$$V = \frac{16}{9}Sin^4\frac{t}{2}f(\theta) \tag{3.4.11}$$

where $f(\theta) = Cosh\theta$ & $Sin\theta$ for Bianchi type-*VIII* and *IX*

respectively.

The average scale factor

$$a(t) = V^{\frac{1}{3}} = \left[\frac{16}{9}Sin^4\frac{t}{2}\right]^{\frac{1}{3}}[f(\theta)]^{\frac{1}{3}} \tag{3.4.12}$$

The expression for expansion scalar θ and the shear σ for the models (3.3.18) and (3.3.25) are given by

$$\theta = 3H = 2Cot\frac{t}{2} \tag{3.4.13}$$

$$\sigma^2 = \frac{14}{9}Cot^2\frac{t}{2} \tag{3.4.14}$$

The deceleration parameter q for the models (3.3.18) and (3.3.25) is given by

$$q = \frac{d}{dt}\left(\frac{1}{H}\right) - 1 = \frac{3}{4}Sec^2\frac{t}{2} - 1 \tag{3.4.15}$$

From equation (3.4.15), we can observe that the deceleration parameter q is always in the range $-1 < q < 0$ and hence they represent accelerating Universes.

The components of the Hubble parameter H_x, H_y & H_z for the models (3.3.18) and (3.3.25) are given by

$$H_x = H_y = \frac{\dot{R}}{R} = \frac{1}{2}Cot\frac{t}{2}, \ H_z = \frac{\dot{S}}{S} = Cot\frac{t}{2}$$

Therefore the generalized mean Hubble parameter (H) is

$$H = \frac{1}{3}\left(H_x + H_y + H_z\right) = \frac{2}{3}Cot\frac{t}{2} \qquad\qquad (3.4.16)$$

The average anisotropy parameter for the models (3.3.18) and (3.3.25) are defined by

$$A_m = \frac{1}{3}\sum_{i=1}^{3}\left(\frac{H_i - H}{H}\right)^2 = \frac{1}{8}\ , \qquad \text{where } i = 1,2,3 \qquad (3.4.17)$$

Red shift

$$Z = \frac{1}{a(t)} - 1 = \left[\frac{16}{9}Sin^4\frac{t}{2}\right]^{-\frac{1}{3}}[f(\theta)]^{-\frac{1}{3}} - 1 \qquad\qquad (3.4.18)$$

The jerk parameter is

$$j(t) = \frac{1}{H^3}\frac{\ddot{a}}{a} = -\frac{1}{8} - \frac{9}{8}Tan^2\frac{t}{2} \qquad\qquad (3.4.19)$$

The tensor of rotation $w_{ij} = u_{i,j} - u_{j,i}$ is identically zero and hence this Universe is non-rotational.

3.5 Conclusions

In this chapter, we have presented spatially homogeneous and anisotropic Bianchi type- *II, VIII* and *IX* cosmological models filled with perfect fluid in the framework of $f(R,T)$ gravity proposed by Harko et al. (2011) with a general choice of the function $f(R,T) = f_1(R) + f_2(T)$ where $f_1(R) = \lambda_1 R$ & $f_2(T) = \lambda_2 T$.

The following are the observations and conclusions:

	Bianchi type-*II* Universe	Bianchi type-*VIII* and *IX* Universes
Singularity	No singularity at $t = \dfrac{-b}{a}$ for $0 < n < 2$	No singularity at $t = 0$
Spatial volume V	At $t = \frac{-b}{a}$, the spatial volume vanishes and increases continuously with time for $0 < (n \neq 1) < 2$. This shows that at the initial epoch the Universe starts with zero volume and expands continuously with time.	Increases with time

Expansion scalar θ	Decreases with the increase of time and diverges at $t = \frac{-b}{a}$	Decreases with the increase of time
Shear scalar σ	Decreases with the increase of time and diverges at $t = \frac{-b}{a}$	Decreases with the increase of time
Hubble parameter H	Decreases with the increase of time and diverges at $t = \frac{-b}{a}$	Decreases with the increase of time
Energy density ρ and pressure p	Tends to infinity as $t \to 0$ and tends to zero as $t \to \infty$	1.Tends to unity as $t \to 0$ and tends to zero as $t \to \infty$. 2. At $t = \pi$, $\rho = p$ and hence they represent Zeldovich fluid

		distribution in general relativity.
Deceleration parameter q	1. For $1 < n < 2$, $-1 < q < 0$ and hence this Universe represents accelerating power law expansion. 2. For $n > 2$, $q < -1$ and hence this Universe expands super exponentially.	q is always in the range $-1 < q < 0$ and hence they represent accelerating Universes.
Mean anisotropy parameter A_m	Since $A_m \neq 0$, which indicates that this model is always anisotropic.	Since $A_m \neq 0$, which indicates that these models are always anisotropic.
Jerk Parameter j	$\dfrac{4(n-1)(7n-10)}{(n+2)^2}$	$-\dfrac{1}{8} - \dfrac{9}{8}\tan^2 \gamma t$

All the models presented here are non singular, anisotropic, non-rotating, expanding and also accelerating. Interestingly, for $\lambda_1 = 1 \,\&\, \lambda_2 = 2\mu$ (i.e., for $f(R,T)=R+2\mu T$), the present models reduce to the Bianchi type-*II*, *VIII* and *IX* perfect fluid cosmological models as obtained by Rao et al. (2014a).

Chapter-4

Five dimensional Kaluza-Klein and Kantowski-Sachs perfect fluid models in $f(R,T)$ gravity*

* Work presented, in this chapter is covered by the following publications:

➢ Higher dimensional perfect fluid cosmological models in $f(R, T)$ gravity: Prespacetime Journal, 5, 1389 (2014).

➢ Kantowski-Sachs cosmological model in in $f(R, T)$ theory of gravity: The African Review of Physics 10, 0019 (2015).

4.1 Introduction

Recent data from several sources, i.e., Supernovae type Ia, cosmic microwave background, large scale structure etc provide evidences about Universe accelerated expansion in current regime (Perlmutter et al. 1999; Reiss et al. 2007; Komatsu et al. 2011). This behavior of the Universe is assumed to be due to the presence of dark energy which holds large negative pressure. The proposal of extended theories of gravity is one of the attractive approach to deal with mysterious nature of dark energy. In this perspective, Capozziello and Laurentis (2011) discussed several criteria for any modified theory, especially scalar-tensor theories of gravity and $f(R)$ theory, to be well-consistent with observational cosmology. Nojiri and Odintsov (2011) reviewed Universe cosmic inflation by means of various $f(R)$ dark energy models. The $f(R,T)$ gravity (Harko et al. 2011) is the modification of $f(R)$ theory, where T dependence is induced by quantum effects or exotic non-ideal matter configurations.

Houndjo (2012) has developed the cosmological reconstruction of $f(R,T)$ gravity with $f(R,T) = f_1(R) + f_2(T)$

and discussed the transition of matter dominated phase to an accelerated phase. Shabani and Farhoudi (2013) have discussed cosmological models in this theory by choosing $f(R,T) = f_1(R) + f_2(T)$. Rao and Neelima (2013c) have discussed perfect fluid non-static plane symmetric Universe in this modified theory. Recently, Rao and Papa Rao (2015b) have investigated Bianchi type-V string cosmological models in $f(R,T)$ gravity. Mishra et al. (2015) have obtained non-static cosmological model in $f(R,T)$ gravity.

The study of higher-dimensional space-time is important because of the underlying idea that the cosmos at its early stages of evolution of the Universe might have had a higher dimensional era. Higher dimensional theories of Kaluza-Klein type have been considered to study some aspects of early Universe. This fact attracted many earlier researchers (Chodos and Detweiler 1980; Freund 1982; Sahdev 1984; Shafi and Wetterich 1984; Witten 1984; Appelquist et al. 1987) in this field of higher dimension. Appelquist and Chodos (1983), Randjbar-Daemi et al. (1984) claimed through solution of the field equations that there is an expansion of four-dimensional space-time while fifth dimension contracts to the unobservable

Plankian length scale or remains constant as needed for the real Universe. Solutions of field equations in higher dimensional space-time are believed to be of physical relevance possibly at the early times before the Universe has undergone compactification transitions. Further, Marciano (1984) has suggested that the experimental observations of fundamental constants with varying time could produce the evidence of extra dimensions. Rahaman (2002), Rahaman et al. (2003a, b, c), Mohanty et al. (2007), Chaubey and Shukla (2013) and Naidu et al. (2013) are some of the authors who have discussed various aspects of higher dimensional cosmological models in different theories of gravitation. Rao and Sireesha (2012b) have obtained a higher dimensional string cosmological model in scalar-tensor theory of gravitation. Rao et al. (2012b) have discussed Kaluza-Klein radiating model in a general scalar-tensor theory. Reddy et al. (2012c) have obtained Kaluza-Klein dark energy cosmological model in Saez-Ballester theory of gravitation. Reddy et al. (2012b) have investigated a five dimensional Kaluza-Klein space-time in the presence of a perfect fluid source in $f(R,T)$ theory of gravitation with negative constant deceleration parameter. Biswal et al. (2015)

have discussed Kaluza-Klein cosmological model in $f(R,T)$ gravity with domain walls.

The Kantowski-Sachs cosmologies have two symmetry properties, the spherical symmetry and the invariance under spatial translations. The vacuum solution for this line element is equivalent to the inner Schwarzschild space time and exact solutions were also found in the presence of some matter fields for homogeneous cosmological models. Kantowski-Sachs (1966) provided solutions for dust space times but later on Kantowski-Sachs geometries with other matter sources were found, such as perfect fluid (Collins 1977), scalar fields (Barrow and Dabrowski 1997), anisotropic fluid (Gergely 1999) and exotic fluid (Gergely 2002) models. Friedmann-Robertson-Walker (FRW) models, being spatially homogeneous and isotropic in nature, are best suited for the representation of large scale structure of the present Universe. However, it is believed that the early Universe may not have been exactly uniform in its expansion phase. Thus, the models with anisotropic background seemed the most suitable to describe the early stages of the Universe. Kantowski-Sachs space time is the simplest model with anisotropic background. Wang (2005) investigated string cosmological models with bulk viscosity in Kantowski-Sachs space time. Chaubey (2012) investigated Kantowski-Sachs model field with perfect fluid in Lyra's geometry. Rao and Prasanthi (2015a,b) have discussed Kantowski-Sachs cosmological models in Lyra's geometry and $f(R,T)$ modified theory of gravity.

Motivated by the above investigations, we discuss, in this chapter, a spatially homogeneous, anisotropic five dimensional Kaluza-Klein and Kantowski-Sachs cosmological models in the presence of perfect fluid source in the framework of $f(R,T)$ gravity proposed by Harko et al. (2011) with a general choice of the function $f(R,T) = f_1(R) + f_2(T)$.

This chapter is arranged as follows: In section 4.2, we derived the field equations of $f(R,T)$ theory of gravity with the help of five dimensional Kaluza-Klein metric, we obtained solution of the field equations and also discussed some features of the models with conclusions. In section 4.3, we have presented field equations of $f(R,T)$ theory of gravity and their solution with the help of a spatially homogeneous Kantowski-Sachs metric. Some important features with conclusions of the model are discussed in the last section 4.4.

4.2 Five Dimensional Kaluza-Klein Model

4.2.1 Metric and field equations:

We consider spatially homogeneous five dimensional Kaluza-Klein metric in the form

$$ds^2 = dt^2 - A^2(t)(dx^2 + dy^2 + dz^2) - B^2(t)dm^2 \qquad (4.2.1)$$

where A and B metric potentials and functions of 't' only .

The matter tensor for perfect fluid is

$$T^i_j = (\rho + p)u^i u_j - \delta^i_j p \qquad (4.2.2)$$

here u^i is the four velocity vector in commoving coordinates, which satisfies the condition $u^i u_i = 1$. Here ρ and p are the energy density and pressure of the perfect fluid respectively.

A systematic derivation of filed equations of $f(R,T)$ gravity is presented in the previous two chapters, however for ready reference they are given by

$$f_1'(R)R_{ij} - \frac{1}{2}f_1(R)g_{ij} = 8\pi T_{ij} + f_2'(T)T_{ij} + [f_2'(T)p + \tfrac{1}{2}f_2(T)]g_{ij} \qquad (4.2.3)$$

where T_{ij} is the energy momentum tensor and T is the trace of the energy momentum tensor of the matter. Here the prime denotes differentiation w.r.t argument.

Now, we consider $f_1(R) = \lambda_1 R$ and $f_2(T) = \lambda_2 T$, so that the field equations (4.2.3) reduces to

$$R_j^i - \frac{1}{2}\delta_j^i R \equiv \left(\frac{8\pi + \lambda_2}{\lambda_1}\right)T_j^i + \frac{\lambda_2}{\lambda_1}\left(p + \frac{T}{2}\right)\delta_j^i \qquad (4.2.4)$$

where λ_1 and λ_2 are arbitrary constants.

4.2.2 Solution of the field equations:

Now with the help of (4.2.2), the field equations (4.2.4) for the metric (4.2.1) can be written as

$$2\frac{\ddot{A}}{A} + \frac{\dot{A}^2}{A^2} + 2\frac{\dot{A}\dot{B}}{AB} + \frac{\ddot{B}}{B} = \left(\frac{8\pi + 2\lambda_2}{\lambda_1}\right)p - \frac{\lambda_2}{2\lambda_1}\rho \qquad (4.2.5)$$

$$3\frac{\dot{A}^2}{A^2} + 3\frac{\dot{A}\dot{B}}{AB} = -\left(\frac{16\pi + 3\lambda_2}{2\lambda_1}\right)\rho + \frac{\lambda_2}{\lambda_1}p \qquad (4.2.6)$$

$$3\frac{\ddot{A}}{A} + 3\frac{\dot{A}^2}{A^2} = \left(\frac{8\pi + 2\lambda_2}{\lambda_1}\right)p - \frac{\lambda_2}{2\lambda_1}\rho \qquad (4.2.7)$$

here the overhead dot denotes differentiation with respect to t.

The field equations (4.2.5) to (4.2.7) are three independent equations in four unknowns A, B, ρ and p. In order to get a deterministic solution we take the following plausible physical condition, the shear scalar σ is proportional to scalar expansion θ, which leads to the linear relationship between the metric potentials A and B, i.e.,

$$A = B^n \tag{4.2.8}$$

where $n \neq 0$ is a constant.

From equations (4.2.5), (4.2.7) and (4.2.8), we get

$$B = \left[(3n+1)(k_1 t + k_2)\right]^{\frac{1}{3n+1}} \tag{4.2.9}$$

where k_1 and k_2 are integrating constants.

From equations (4.2.8) and (4.2.9), we get

$$A = \left[(3n+1)(k_1 t + k_2)\right]^{\frac{n}{3n+1}} \tag{4.2.10}$$

From equations (4.2.5) to (4.2.7), (4.2.9) and (4.2.10) we get

the energy density

$$\rho = \frac{-6\lambda_1(8\pi + 3\lambda_2)n(n+1)}{(16\pi + 5\lambda_2)(8\pi + \lambda_2)}\left(\frac{k_1}{(3n+1)(k_1 t + k_2)}\right)^2 \tag{4.2.11}$$

the total pressure

$$p = \frac{-12\lambda_1(4\pi+\lambda_2)n(n+1)}{(16\pi+5\lambda_2)(8\pi+\lambda_2)}\left(\frac{k_1}{(3n+1)(k_1 t+k_2)}\right)^2.$$
(4.2.12)

Since $p<0$ for $n>0$, it shows that the perfect fluid behave like a phantom-type dark energy. So we can conclude that the perfect fluid may be a source of early dark energy due to negative pressure, since energy conditions are violated. But for $-1<n<0$, $p>0$ & $\rho>0$ and hence the equations (4.2.11) and (4.2.12) satisfy the energy conditions.

The metric (4.2.1) can now be written as

$$ds^2 = dt^2 - \left[(3n+1)(k_1 t+k_2)\right]^{\frac{2n}{3n+1}}\left(dx^2+dy^2+dz^2\right)$$
$$- \left[(3n+1)(k_1 t+k_2)\right]^{\frac{2}{3n+1}}dm^2$$
(4.2.13)

Thus the metric (4.2.13) together with (4.2.11) and (4.2.12) constitutes a five-dimensional Kaluza-Klein perfect fluid cosmological model in $f(R,T)$ theory of gravity.

For particular values of $\lambda_1 = 1$ & $\lambda_2 = 2\lambda$, $f(R,T)=R+2\lambda T$. Then the total pressure and energy density will become

$$p = \frac{-6(2\pi+\lambda)n(n+1)}{(8\pi+5\lambda)(4\pi+\lambda)}\left(\frac{k_1}{(3n+1)(k_1 t+k_2)}\right)^2$$
(4.2.14)

$$\rho = \frac{-3(4\pi + 3\lambda)n(n+1)}{(8\pi + 5\lambda)(4\pi + \lambda)}\left(\frac{k_1}{(3n+1)(k_1 t + k_2)}\right)^2 \qquad (4.2.15)$$

For the above particular values of $\lambda_1 \& \lambda_2$, the metric (4.2.13) together with (4.2.14) and (4.2.15) represents Kaluza–Klein perfect fluid cosmological model in $f(R,T)$ gravity with $f(R,T)=R+2\lambda T$, which is more general than the model investigated by Reddy et al. (2012b). Also, since we have obtained this model without taking the additional condition of special law of variation of Hubble parameter proposed by Bermann (1983), this additional condition taken by Reddy et al. (2012b) is superfluous.

Isotropic cosmological model:

For $n=1,$ the metric (4.2.13) together with (4.2.11) and (4.2.12) constitutes a five-dimensional Kaluza-Klein perfect fluid isotropic cosmological model in $f(R,T)$ theory of gravity.

4.2.3 Some important features and conclusions:

The volume element of the model (4.2.13) is given by

$$V = (g)^{1/2} = (3n+1)\left(k_1 t + k_2\right) \qquad (4.2.16)$$

Average scale factor

$$a(t) = V^{\frac{1}{4}} = \left[(3n+1)(k_1 t + k_2) \right]^{\frac{1}{4}} \tag{4.2.17}$$

The expression for the expansion scalar θ is given by

$$\theta = u^i_{,i} = \frac{k_1}{k_1 t + k_2} \tag{4.2.18}$$

and the shear σ is given by

$$\sigma^2 = \frac{1}{2} \sigma^{ij} \sigma_{ij} = \frac{4}{9} \left(\frac{k_1}{k_1 t + k_2} \right)^2 \tag{4.2.19}$$

The deceleration parameter q is given by

$$q = (-3\theta^{-2})(\theta_{,i} u^i + \frac{1}{3}\theta^2) = 2 \tag{4.2.20}$$

The Hubble's parameter H is given by

$$H = \frac{\theta}{4} = \frac{k_1}{4(k_1 t + k_2)} \tag{4.2.21}$$

The overall density parameter Ω is given by

$$\Omega = \frac{\rho}{3H^2} = \frac{-3\lambda_1}{(2n+1)^2} \left(\frac{n+1}{8\pi + 2\lambda_2} + \frac{n}{8\pi + \lambda_2} \right) \tag{4.2.22}$$

The mean anisotropy parameter A_m is given by

$$A_m = \frac{1}{4} \sum_{i=1}^{4} \left(\frac{H_i - H}{H} \right)^2 = \frac{3(n-1)^2}{(3n+1)^2} \tag{4.2.23}$$

where $\Delta H_i = H_i - H \quad (i = 1,2,3,4)$

Red shift

$$Z = \frac{1}{a} - 1 \ = \left[(3n+1)(k_1 t + k_2)\right]^{\frac{-1}{4}} - 1 \qquad (4.2.24)$$

Jerk parameter

$$j = \frac{1}{H^3}\left(\frac{\dddot{a}}{a}\right) = 10 \qquad (4.2.25)$$

Luminosity parameter $d_L = r_1(1+z)$, where $r_1 = \int_t^{t_0} \frac{1}{a(t)} dt$

$$d_L = \frac{3(3n+1)^{\frac{-2}{3}} (k_1 t + k_2)^{\frac{1}{3}}}{2k_1}\left[\left(\frac{k_1 t_0 + k_2}{k_1 t + k_2}\right)^{\frac{2}{3}} - 1\right] \qquad (4.2.26)$$

The distance modulus

$$D(Z) = 5\log d_L + 25$$

i.e. $\quad D(Z) = 5\log\left[\frac{H_0^{-1}(1+z)}{(k+3)}\left[1 - (1+z)^{-(k+3)}\right]\right] + 25 \qquad (4.2.27)$

Conclusions:

Here, we have presented spatially homogeneous and anisotropic five dimensional Kaluza-Klein cosmological models filled with perfect fluid in the framework of $f(R,T)$ gravity proposed by Harko et al. (2011) with an appropriate choice of a function $f(R,T) = f_1(R) + f_2(T)$. For Kaluza-Klein cosmological model, we observe that at $t = \frac{-k_2}{k_1}$, the spatial volume vanishes and increases continuously with time while all other parameters diverge. This shows that at the initial epoch the Universe starts with zero volume and expands continuously approaching to infinite volume. At $t = \frac{-k_2}{k_1}$, the expansion scalar θ , the shear scalar σ and Hubble parameter H diverges. Also the model has no singularity for $n > 0$. From (4.2.11) and (4.2.12), we can see that matter pressure and energy density will vanish with the increase of time. Also from (4.4.5) we can observe that the deceleration parameter is always positive and hence the Universe decelerates in standard way and which is similar to the five dimensional Kaluza-Klein $f(R)$ gravity as discussed by Aghmohammadi et al. (2009) and Huang et al. (2010). However, in spite of the fact that the Universe, in this case, decelerates in the standard way it will accelerate in finite time due to cosmic re collapse where the Universe in turns inflates "decelerates and then accelerates" (Nojiri and Odintsov 2003). Thus, even though five dimensional $f(R,T)$ theory of gravity decelerates in a standard way, it will accelerate in finite time, thus establishing consistency with the present day accelerated expansion of the Universe. It may be mentioned here that in five dimensional $f(R)$ and $f(R,T)$ theories of gravity both expansion and contraction of the extra dimension could result in the present accelerated expansion of other spatial dimensions. It may also be noted that Kaluza-Klein model represents the cosmos in its early stage of evolution. From (4.2.23), we can observe that $A_m \neq 0$, which indicates that the Kaluza-Klein cosmological model is anisotropic for $n \neq 1$. For $n = 1$, the metric (4.2.13) together with (4.2.11) and (4.2.12) constitutes a five-dimensional Kaluza-Klein isotropic perfect fluid cosmological model in $f(R,T)$ theory of gravity. These exact models represent not only the early stages of evolution but also the present Universe.

4.3 Kantowski-Sachs Model

4.3.1 Metric and solution of the field equations:

We consider a spatially homogeneous Kantowski-Sachs metric of the form

$$ds^2 = dt^2 - A^2 dr^2 - B^2\left(d\theta^2 + Sin^2\theta\, d\varphi^2\right) \tag{4.3.1}$$

where A and B are the functions of time t only.

Now with the help of (4.2.2), the field equations (4.2.4) for the metric (4.3.1) can be written as

$$2\frac{\ddot{B}}{B}+\frac{\dot{B}^2}{B^2}+\frac{1}{B^2}=\left(\frac{16\pi+3\lambda_2}{2\lambda_1}\right)p-\frac{\lambda_2}{2\lambda_1}\rho \qquad (4.3.2)$$

$$\frac{\ddot{A}}{A}+\frac{\ddot{B}}{B}+\frac{\dot{A}\dot{B}}{AB}=\left(\frac{16\pi+3\lambda_2}{2\lambda_1}\right)p-\frac{\lambda_2}{2\lambda_1}\rho \qquad (4.3.3)$$

$$\frac{\ddot{A}}{A}+\frac{\ddot{B}}{B}+\frac{\dot{A}\dot{B}}{AB}=\left(\frac{16\pi+3\lambda_2}{2\lambda_1}\right)p-\frac{\lambda_2}{2\lambda_1}\rho \qquad (4.3.4)$$

$$\frac{\dot{B}^2}{B^2}+2\frac{\dot{A}\dot{B}}{AB}+\frac{1}{B^2}=-\left(\frac{16\pi+3\lambda_2}{2\lambda_1}\right)\rho+\frac{\lambda_2}{2\lambda_1}p \qquad (4.3.5)$$

here the dot denotes differentiation with respect to time t.

The equations (4.3.2) to (4.3.5) is a system of three independent equations in four unknowns A, B, p and ρ. In order to get a deterministic solution we take the following plausible physical condition, the shear scalar σ is proportional to scalar expansion θ, which leads to the linear relationship between the metric potentials A and B, i.e.,

$$B = A^n \qquad (4.3.6)$$

where $n \neq 0$ is a constant.

From equations (4.3.2), (4.3.3) and (4.3.6), we get

$$A = \left[n(k_1 t + k_2)\right]^{\frac{1}{n}} \qquad (4.3.7)$$

where $k_1 = \left(\dfrac{1}{1-n^2}\right)^{1/2}$, $n \neq \pm 1$ and k_2 is an integration constant.

From equations (4.3.6) and (4.3.7), we get

$$B = \left[n(k_1 t + k_2)\right] \tag{4.3.8}$$

From equations (4.3.2) to (4.3.5) and (4.3.7), (4.3.8), we get

the energy density as

$$\rho = \frac{\lambda_1}{2n^2\left(n^2 - 1\right)\left(k_1 t + k_2\right)^2}\left(\frac{\lambda_2(3n+1) + 8\pi(2n+1)}{(4\pi + \lambda_2)(8\pi + \lambda_2)}\right) \tag{4.3.9}$$

the total pressure as

$$p = \frac{\lambda_1}{2n^2\left(n^2 - 1\right)\left(k_1 t + k_2\right)^2}\left(\frac{\lambda_2(n-1) - 8\pi}{(4\pi + \lambda_2)(8\pi + \lambda_2)}\right) \tag{4.3.10}$$

From (4.3.7) and (4.3.8) the metric (4.3.1) can be written as

$$ds^2 = dt^2 - \left[n(k_1 t + k_2)\right]^{\frac{2}{n}} dr^2 - \left[n(k_1 t + k_2)\right]^2\left(d\theta^2 + Sin^2\theta\, d\phi^2\right)$$

$$\tag{4.3.11}$$

Thus the metric (4.3.11) together with (4.3.9) and (4.3.10) constitutes Kantowski-Sachs perfect fluid cosmological model in $f(R,T)$ gravity, where $f(R,T) = \lambda_1 R + \lambda_2 T$.

For particular values of $\lambda_1 = 1 \,\&\, \lambda_2 = 2\lambda$, $f(R,T) = R + 2\lambda T$, then the energy density and the total pressure will become

$$\rho = \frac{1}{4n^2\left(n^2-1\right)\left(k_1 t + k_2\right)^2}\left(\frac{4\pi(2n+1)+\lambda(3n+1)}{(2\pi+\lambda)(4\pi+\lambda)}\right) \qquad (4.3.12)$$

$$p = \frac{-1}{4n^2\left(n^2-1\right)\left(k_1 t + k_2\right)^2}\left(\frac{4\pi+\lambda(1-n)}{(2\pi+\lambda)(4\pi+\lambda)}\right) \qquad (4.3.13)$$

For $-1 < n < 1$ and for $1 < n < \infty$ with $\lambda > \dfrac{4\pi}{n-1}$, $p > 0$ and $\rho > 0$.

Hence the equations (4.3.12) and (4.3.13) satisfy the energy conditions. Also $p < 0$, for $-\infty < n < 1$ as well as for $1 < n < \infty$ with $\lambda < \dfrac{4\pi}{n-1}$, it shows that the perfect fluid behave like a phantom-type dark energy. So, we can conclude that the perfect fluid may be a source of early dark energy due to negative pressure, since energy conditions are violated.

For the above particular values of $\lambda_1 \& \lambda_2$, the metric (4.3.11) together with (4.3.12) and (4.3.13) represents Kantowski-Sachs perfect fluid cosmological model in $f(R,T)$ gravity with $f(R,T) = R + 2\lambda T$, which is more general and the additional condition, special law of variation of Hubble parameter proposed by Bermann (1983), taken by Samanta (2013) is superfluous.

4.3.2 Some important features and conclusions:

The volume element of the model (4.3.9) is given by

$$V = (-g)^{1/2} = \left[n(k_1 t + k_2)\right]^{\frac{2n+1}{n}} Sin\theta \qquad (4.3.14)$$

we can observe that the spatial volume is increasing with time.

Average scale factor

$$a(t) = V^{1/3} = \left[n(k_1 t + k_2)\right]^{\frac{2n+1}{3n}} Sin^{1/3}\theta \qquad (4.3.15)$$

The expression for the expansion scalar θ is given by

$$\theta = u^i{}_{,i} = \left(\frac{2n+1}{n}\right)\frac{k_1}{k_1 t + k_2} \qquad (4.3.16)$$

and the shear σ is given by

$$\sigma^2 = \frac{1}{2}\sigma^{ij}\sigma_{ij} = \frac{7}{18}\left(\frac{(2n+1)k_1}{n(k_1 t + k_2)}\right)^2 \qquad (4.3.17)$$

The deceleration parameter q is given by

$$q = -\frac{a\ddot{a}}{\dot{a}^2} = \frac{n-1}{2n+1} \qquad (4.3.18)$$

The Hubble's parameter H is given by

$$H = \frac{\theta}{3} = \frac{(2n+1)k_1}{3n(k_1 t + k_2)} \qquad (4.3.19)$$

The overall density parameter Ω is given by

$$\Omega = \frac{\rho}{3H^2} = \frac{-3\lambda_1}{(2n+1)^2}\left(\frac{n+1}{8\pi+2\lambda_2} + \frac{n}{8\pi+\lambda_2}\right) \qquad (4.3.20)$$

The mean anisotropy parameter A_m is given by

$$A_m = \frac{1}{3}\sum_{i=1}^{3}\left(\frac{H_i - H}{H}\right)^2 = 2\left(\frac{n-1}{2n+1}\right)^2 \qquad (4.3.21)$$

where $\Delta H_i = H_i - H$ $(i=1,2,3)$

Red shift

$$Z = \frac{1}{a} - 1 = \left[n(k_1 t + k_2)\right]^{-\left(\frac{2n+1}{3n}\right)} Sin^{1/3}\theta - 1 \qquad (4.3.22)$$

Jerk parameter

$$j = \frac{1}{H^3}\left(\frac{\ddot{a}}{a}\right) = \frac{(4n-1)(n-1)}{(2n+1)^2} \qquad (4.3.23)$$

From equation (4.4.22) it can be observed that for $n = -0.087658$ the jerk parameter value overlap with the value j $\approx$ 2.16, which is obtained from the three kinematical data sets: the gold sample of type Ia supernovae, the SNIa data from the SNLS project and the X-ray galaxy cluster distance measurements.

Luminosity parameter $d_L = r_1(1+z)$, where $r_1 = \int_t^{t_0} \frac{1}{a(t)}dt$

$$d_L = \frac{3(n)^{\frac{-(n+2)}{3n}}}{(n-1)k_1 Sin^{2/3}\theta}\left((k_1 t_0 + k_2)^{\frac{n-1}{3n}}(k_1 t + k_2)^{\frac{-(2n+1)}{3n}} - (k_1 t + k_2)^{\frac{-(n+2)}{3n}}\right) \qquad (4.3.24)$$

The distance modulus $D(Z) = 5\log d_L + 25$

$$\text{i.e.} \quad D(Z) = 5\log\left[\frac{H_0^{-1}(1+z)}{(k+3)}\left[1-(1+z)^{-(k+3)}\right]\right] + 25 \qquad (4.3.25)$$

The tensor of rotation $w_{ij} = u_{i,j} - u_{j,i}$ is identically zero and hence this Universe is non-rotational.

Conclusions:

Here, we have presented spatially homogeneous and anisotropic Kantowski-Sachs cosmological model filled with perfect fluid in the framework of $f(R,T)$ gravity proposed by Harko et al. (2011) have been obtained with an appropriate choice of a function $f(R,T) = f_1(R) + f_2(T)$. We observe that at $t = -k_2\sqrt{1-n^2}$, the spatial volume vanishes and increases continuously with time for $n > -0.5$. This shows that at the initial epoch the Universe starts with zero volume and expands continuously with time. Also the model has no singularity for $n > 0$. The expansion scalar θ, shear scalar σ and the Hubble parameter H decreases with the increase of time and diverges at $t = -k_2\sqrt{1-n^2}$. From (4.3.9) and (4.3.10), we can see that energy density and matter pressure will vanish with the increase of time. From (4.3.18), we can observe that the deceleration parameter is negative for $-0.5 < n < 1$ and hence it represents an accelerating Universe. Since $A_m \neq 0$, this indicates that this model is always anisotropic. The model presented here is anisotropic, non-rotating, expanding and also accelerating. Interestingly, for particular values of λ_1 & λ_2, the metric (4.3.11) together with (4.3.12) and (4.3.13) represents Kantowski-Sachs perfect fluid cosmological model in $f(R,T)$ gravity with $f(R,T)=R+2\lambda T$, which is more general than the model investigated by Samanta (2013).

Chapter-5

Tilted Bianchi type-*I* model in Brans-Dicke theory and tilted Bianchi type-*II, VIII, IX* models in Lyra manifold*

* Work presented, in this chapter, is covered by the following publications:

➤ A Tilted Cosmological Model in Brans-Dicke Theory of Gravitation: Prespacetime Journal, 6, 777 (2015).

➤ Tilted Bianchi Cosmological Models in Lyra Manifold Communicated to The African review of physics.

5.1 Introduction

The tilted cosmological models, in which the fluid vector is not normal to the hyper surface of homogeneity, are more complicated than those of non-tilted one. These tilted cosmological models help to study the effect of large scale peculiar velocity field relative to cosmic microwave background radiation. Further tilted cosmological models help to predict the growth of the inhomogeneous models and also it establish the relationship with the observed large scale structure. The general dynamics of tilted cosmological models have been examined by King and Ellis (1973), Ellis and King (1974) and Collins and Ellis (1979). Dunn and Tupper (1978) have investigated tilted Bianchi type-I cosmological model for perfect fluid. Bali and Sharma (2002) have studied tilted Bianchi type-I cosmological model with dust fluid. Pradhan and Rai (2003) revisited tilted Bianchi type-I cosmological models filled with disordered radiation in general relativity investigated by Bali and Meena (2002). Sahu (2010) has obtained tilted Bianchi type-VI_0 cosmological model in Saez-Ballester scalar-tensor theory of gravitation. Anita (2012) has studied tilted Bianchi type-IX dust

115

fluid cosmological model in general relativity. Sahu and Kumar (2013) have obtained tilted Bianchi type-I cosmological model in Lyra geometry. Bali and Kumawat (2015) have investigated LRS Bianchi type-II tilted barotropic fluid cosmological model with heat conduction in general relativity. Recently, Sahu et al. (2015a) have studied tilted Bianchi type-VI_0 wet dark fluid cosmological model in general relativity. Sahu et al. (2015b) have obtained tilted Bianchi type-III cosmological models in Lyra geometry. Pawar and Dagwal (2015) have investigated tilted kasner type cosmological models in Brans-Dicke theory of gravitation.

Biachi space-times play a vital role in understanding and description of the early stages of evolution of the Universe. Moreover, from the theoretical point of view anisotropic Universes have a greater generality than isotropic models. Space-times admitting a three-parameter group of automorphisms are important in the discussion of cosmological models. Bianchi has shown that there are only nine distinct sets of structure constants for groups of this type so that the algebra may be easily used to classify homogeneous space-times. Thus, Bianchi space-times admit a three parameter group of motions and hence they have only a manageable number of degrees of freedom. A complete list of all exact solutions of Einstein's equations for the Bianchi type *I-IX* with perfect fluid matter is given by Kramer et al. (1980).

From last few decades there has been a lot of interest in scalar tensor theories of gravitation. Noteworthy among them Sen theory is the scalar tensor theory proposed by Sen (1957) based on Lyra (1951) geometry and Brans-Dicke scalar-tensor theory of gravitation. A detailed description of these theories were presented in chapter-1.

Reddy (2003) has obtained a string cosmological model in Brans-Dicke theory of gravitation. Chi-Yi Chen and You-Gen Shen (2006) have studied holographic principle of black holes in Brans-Dicke theory. Reddy et al. (2007) have studied a cosmological model with negative constant deceleration parameter in Brans-Dicke theory. Adhav et al. (2009) have discussed plane symmetric vacuum Bianchi type-*III* cosmological model in Brans-Dicke theory. Rao and Vijaya Santhi (2011, 2012b, c) have discussed different aspects of Bianchi type-*II*, *VIII* and *IX* cosmological models in Brans-Dicke theory of gravitation. Naidu et al. (2013) have studied a five dimensional Kaluza-Klein bulk viscous string cosmological model in Brans-Dicke scalar-tensor theory of gravitation. Wankhade and Sancheti (2014) have discussed Bianchi type-*II*, *VIII* and *IX* cosmological models with magnetized anisotropic dark energy in Brans-Dicke theory of gravitation. Very recently, Rao et al. (2015d) have investigated FRW holographic dark energy cosmological model in Brans-Dicke theory of gravitation. Rao and Jayasudha (2015a) have studied Bianchi type-V dark energy model in Brans-Dicke theory. Rao and Jayasudha (2015b) have obtained five dimensional spherically symmetric cosmological model in Brans-Dicke theory of gravitation. Rao and Jayasudha (2015c) have discussed Bianchi type-VI_0 anisotropic dark energy model in Brans-Dicke theory of gravitation.

Halford (1972) has shown that the scalar-tensor treatment based on Lyra's geometry predicts the same effects as in general relativity. Several authors have studied cosmological models within the framework of Lyra geometry with a constant gauge vector in the time direction. Rahaman et al. (2005) have investigated cosmological models with negative constant deceleration parameter in Lyra geometry. Rao and Vinutha (2009) have studied axially symmetric cosmological models in a scalar tensor theory of gravitation based on Lyra geometry. Bali et al. (2012) have studied Bianchi type-*IX* barotropic fluid model with time-dependent displacement vector in Lyra geometry. Samanta and Debata (2012) have studied five dimensional Bianchi type-*I* string cosmological models in Lyra manifold. Singh and Sharma (2014b) have discussed anisotropic dark energy Bianchi type-*II* cosmological model in Lyra geometry. Recently, Katore and Hatkar (2015) have investigated Kaluza-Klein Universe with magnetized anisotropic dark energy in general relativity and Lyra manifold.

Inspired by the above investigations and discussions, in this chapter, we have presented the tilted Bianchi type-*I* stiff fluid cosmological model in Brans-Dicke theory of gravitation and tilted Bianchi type-*II, VIII* and *IX* cosmological models in Sen theory of gravitation based on Lyra geometry.

The paper is organized as follows: In section 5.2, we obtained tilted Bianchi type-*I* model filled with perfect fluid in Brans-Dicke theory of gravitation along with their solution, also we have discussed some important features of the obtained model with conclusions. In section 5.3, we have presented the field equations of Sen theory and their solutions with the help of Bianchi type-*II*, *VIII* and *IX* metrics. Also, some important features of the models with conclusions are discussed.

5.2 Tilted Bianchi Type-*I* Model in Brans-Dicke Theory

5.2.1 Metric and field equations:

We consider Bianchi type-*I* metric in the form

$$ds^2 = -dt^2 + A^2 dx^2 + B^2 dy^2 + C^2 dz^2 \qquad (5.2.1)$$

where A, B, C are functions of cosmic time 't' only.

The energy momentum tensor for a perfect fluid distribution with heat conduction is given by

$$T_j^i = (\rho + p)u_i u^j + pg_i^j + q_i u^j + u_i q^j \qquad (5.2.2)$$

120

together with

$$g_{ij}u^i u^j = -1,$$

(5.2.3)

$$q_i q^i > 0 \text{ and } q_i u^i = 0.$$

(5.2.4)

where p is the pressure and ρ is the energy density of the perfect fluid distribution, q_i is the heat conduction vector orthogonal to u_i. The fluid vector u_i has the components $u^i = \left(\dfrac{\sinh \lambda}{A}, 0, 0, \cosh \lambda \right)$ satisfying (5.2.3) and λ is the tilt angle.

Brans-Dicke theory (1961) of gravitation is well known modified version of Einstein's theory. It is a scalar tensor theory in which the gravitational interaction is mediated by a scalar field ϕ as well as the tensor field g_{ij} of Einstein's theory. In this theory the scalar field ϕ has the dimension of the inverse of the gravitational constant.

$$R_{ij} - \frac{1}{2} R g_{ij} = -8\pi \phi^{-1} T_{ij} - \omega \phi^{-2} \left(\phi_{,i} \phi_{,j} - \frac{1}{2} g_{ij} \phi_{,r} \phi^{,r} \right) - \phi^{-1} (\phi_{i;j} - g_{ij} \phi_{,r}^{\ ,r})$$

(5.2.5)

and $\quad \phi_{,r}^{\ ,r} = 8\pi (3 + 2\omega)^{-1} T$

(5.2.6)

Also, we have energy conservation equation as

$$T^{ij}{}_{;j} = 0 \tag{5.2.7}$$

where T_{ij} is the stress energy tensor of the matter, ω is the dimensionless coupling constant and comma and semi-colon denote partial and covariant differentiation respectively.

The field equations (5.2.5) - (5.2.7) for the metric (5.2.1) with the help of equations (5.2.2) - (5.2.4) can be written as

$$\frac{\ddot{B}}{B} + \frac{\ddot{C}}{C} + \frac{\dot{B}\dot{C}}{BC} + \frac{\omega}{2}\frac{\dot{\phi}^2}{\phi^2} + \frac{\dot{\phi}}{\phi}\left(\frac{\dot{B}}{B} + \frac{\dot{C}}{C}\right) + \frac{\ddot{\phi}}{\phi} = -\frac{8\pi}{\phi}\left\{(\rho + p)\sinh^2\lambda + p + 2q_1\left(\frac{\sinh\lambda}{A}\right)\right\} \tag{5.2.8}$$

$$\frac{\ddot{A}}{A} + \frac{\ddot{C}}{C} + \frac{\dot{A}\dot{C}}{AC} + \frac{\omega}{2}\frac{\dot{\phi}^2}{\phi^2} + \frac{\dot{\phi}}{\phi}\left(\frac{\dot{A}}{A} + \frac{\dot{C}}{C}\right) + \frac{\ddot{\phi}}{\phi} = -\frac{8\pi}{\phi}p \tag{5.2.9}$$

$$\frac{\ddot{B}}{B} + \frac{\ddot{A}}{A} + \frac{\dot{B}\dot{A}}{BA} + \frac{\omega}{2}\frac{\dot{\phi}^2}{\phi^2} + \frac{\dot{\phi}}{\phi}\left(\frac{\dot{B}}{B} + \frac{\dot{A}}{A}\right) + \frac{\ddot{\phi}}{\phi} = -\frac{8\pi}{\phi}p \tag{5.2.10}$$

$$\frac{\dot{A}\dot{B}}{AB} + \frac{\dot{B}\dot{C}}{BC} + \frac{\dot{A}\dot{C}}{AC} - \frac{\omega}{2}\frac{\dot{\phi}^2}{\phi^2} + \frac{\dot{\phi}}{\phi}\left(\frac{\dot{A}}{A} + \frac{\dot{B}}{B} + \frac{\dot{C}}{C}\right) = -\frac{8\pi}{\phi}\left\{-(\rho + p)\cosh^2\lambda + p - 2q_1\left(\frac{\sinh\lambda}{A}\right)\right\} \tag{5.2.11}$$

$$\frac{8\pi}{\phi}\left\{(\rho + p)A\sinh\lambda\cosh\lambda + q_1\cosh\lambda + q_1\left(\frac{\sinh^2\lambda}{\cosh\lambda}\right)\right\} = 0 \tag{5.2.12}$$

$$\ddot{\phi} + \dot{\phi}\left(\frac{\dot{A}}{A} + \frac{\dot{B}}{B} + \frac{\dot{C}}{C}\right) = \frac{8\pi}{3+2\omega}\left\{\rho - 3p - (\rho+p)A\sinh\lambda\cosh\lambda - q_1\cosh\lambda - q_1\left(\frac{\sinh^2\lambda}{\cosh\lambda}\right)\right\}$$

$$(5.2.13)$$

Using transformation $A = e^{\alpha}$, $B = e^{\beta}$, $C = e^{\gamma}$ and $dt = ABC\,dT$, the above field equations can be re-written as

$$\beta'' + \gamma'' - \beta'\alpha' - \beta'\gamma' - \alpha'\gamma' + \frac{\omega}{2}\frac{\phi'^2}{\phi^2} + \frac{\phi''}{\phi} - \frac{\phi'}{\phi}\alpha' =$$
$$\frac{-8\pi}{\phi}\left\{(\rho+p)\sinh^2\lambda + p + 2q_1\left(\frac{\sinh\lambda}{A}\right)\right\}e^{2(\alpha+\beta+\gamma)}$$

$$(5.2.14)$$

$$\alpha'' + \gamma'' - \beta'\alpha' - \beta'\gamma' - \alpha'\gamma' + \frac{\omega}{2}\frac{\phi'^2}{\phi^2} + \frac{\phi''}{\phi} - \frac{\phi'}{\phi}\beta' = \frac{-8\pi}{\phi}p\left(e^{2\alpha+2\beta+2\gamma}\right)$$

$$(5.2.15)$$

$$\alpha'' + \beta'' - \beta'\alpha' - \beta'\gamma' - \alpha'\gamma' + \frac{\omega}{2}\frac{\phi'^2}{\phi^2} + \frac{\phi''}{\phi} - \frac{\phi'}{\phi}\gamma' = \frac{-8\pi}{\phi}p\left(e^{2\alpha+2\beta+2\gamma}\right)$$

$$(5.2.16)$$

$$\alpha'\beta' + \beta'\gamma' + \gamma'\alpha' - \frac{\omega}{2}\frac{\phi'^2}{\phi^2} + \frac{\phi'}{\phi}(\alpha' + \beta' + \gamma') =$$
$$\frac{-8\pi}{\phi}\left\{-(\rho+p)\cosh^2\lambda + p - 2q_1\left(\frac{\sinh\lambda}{e^{\alpha}}\right)\right\}e^{2(\alpha+\beta+\gamma)}$$

$$(5.2.17)$$

$$(\rho + p)e^{\alpha} \sinh\lambda \cosh\lambda + q_1\left(\frac{\cosh 2\lambda}{\cosh\lambda}\right) = 0 \qquad (5.2.18)$$

$$\frac{\phi'' + \phi'(\alpha' + \beta' + \gamma')}{e^{2(\alpha+\beta+\gamma)}} = \frac{8\pi}{3+2\omega}\left\{\rho - 3p - (\rho + p)e^{\alpha}\sinh\lambda\cosh\lambda - q_1\left(\frac{\cosh 2\lambda}{\cosh\lambda}\right)\right\}$$

$$(5.2.19)$$

where prime denotes the differentiation with respect to T.

5.2.2 Solution of the field equations:

The field equations (5.2.14) to (5.2.19) are six independent equations with eight unknowns, so in order to get a deterministic solution we take the following plausible physical conditions:

i. The shear scalar σ is proportional to scalar expansion θ, which leads to the linear relationship between the metric potentials A and B, i.e.

$$A = B^k \Rightarrow \alpha = k\beta \qquad (5.2.20)$$

where $k \neq 0$ is an arbitrary constant.

ii. Stiff fluid condition

$$\text{i.e., } p = \rho. \tag{5.2.21}$$

From equations (5.2.14)-(5.2.17), (5.2.20) and (5.2.21), we get

$$\gamma = c_1 + c_2 e^{\frac{-k_1}{2}T} - k_1 T \tag{5.2.22}$$

$$\beta = c_3 + c_4 e^{-k_1 T} + e^{\frac{-k_1}{2}T} - k_1 T \tag{5.2.23}$$

$$\alpha = k\left(c_3 + c_4 e^{-k_1 T} + e^{\frac{-k_1}{2}T} - k_1 T \right) \tag{5.2.24}$$

and the scalar field

$$\phi = e^{k_1 T + k_2}. \tag{5.2.25}$$

Where c_1, c_2, c_3 and c_4 are integration constants and k_1, k_2 are arbitrary constants.

Thus the metric (5.2.1) can be written as

$$ds^2 = -dt^2 + \exp 2k\left(c_3 + c_4 e^{-k_1 T} + e^{\frac{-k_1}{2}T} - k_1 T \right) dx^2$$

$$+\exp 2\left(c_3 + c_4 e^{-k_1 T} + e^{\frac{-k_1}{2}T} - k_1 T\right) dy^2$$

$$+\exp 2\left(c_1 + c_2 e^{\frac{-k_1}{2}T} - k_1 T\right) dz^2 \tag{5.2.26}$$

From equations (5.2.15) and (5.2.21), we get

$$p = \rho = \frac{k_4 e^{-k_1 T} + k_5 e^{\frac{-3k_1 T}{2}} - k_3 e^{\frac{-k_1 T}{2}} - k_6}{8\pi \exp\left\{2(k+1)\left(c_3 + c_4 e^{-k_1 T} + e^{\frac{-k_1}{2}T} - k_1 T\right) + 2\left(c_1 + k_2 + c_2 e^{\frac{-k_1}{2}T} - (k_1 - 1)T\right)\right\}} \tag{5.2.27}$$

where $\quad k_3 = \dfrac{k_1^2}{4}\left((1-2k) - 2(1+k)c_2\right); \; k_4 = k + 8kc_4 + c_2(k+1);$

$$k_5 = 4k + 2c_4 c_2(k+1); \quad k_6 = -4k + 4 + 2\omega.$$

From equation (5.2.18), we get heat conduction vector as

$$q_1 = \frac{\left\{k_3 e^{\frac{-k_1 T}{2}} + k_6 - k_4 e^{-k_1 T} - k_5 e^{\frac{-3k_1 T}{2}}\right\}}{\exp\left\{c_1 + k_2 + c_3 + c_4 e^{-k_1 T} + (c_2 + 1)e^{\frac{-k_1}{2}T} - (2k_1 + 1)T\right\}} \tanh 2\lambda . \cosh \lambda \tag{5.2.28}$$

From equations (5.2.14), (5.2.18) and (5.2.28), we obtain tilt angle

as

$$\lambda = \frac{1}{2}\cosh^{-1}\left\{\frac{k_4 e^{-k_1 T} + k_5 e^{\frac{-3k_1 T}{2}} - k_3 e^{\frac{-k_1 T}{2}} - k_6}{k_7 e^{-k_1 T} + k_8 e^{\frac{-k_1 T}{2}} + k_9 e^{\frac{-3k_1 T}{2}} - k_{10} e^{-2k_1 T} + 3k_1^2}\right\} \tag{5.2.29}$$

where $k_7 = \dfrac{k_1^2}{12\pi}(12c_4 + 3c_2 + 1)$; $k_8 = \dfrac{k_1^2}{12\pi}(12 + 6c_2)$;

$$k_9 = \frac{k_1^2}{12\pi}(4c_4 + 2c_4 c_2(1 + 2k)) \text{ and } k_{10} = \frac{k_1^2}{3\pi}kc_4^2.$$

5.2.3 Some important features and conclusions:

The volume of the model (5.2.26) is given by

$$V = \exp\left\{(k+1)\left(c_3 + c_4 e^{-k_1 T} + e^{\frac{-k_1}{2}T} - k_1 T\right) + c_1 + c_2 e^{\frac{-k_1}{2}T} - k_1 T\right\} \tag{5.2.30}$$

The expansion scalar θ is given by

$$\theta = \left[\frac{-k_1\left\{(k+1)\left(c_4 e^{-k_1 T} + \frac{k_1}{2}e^{\frac{-k_1}{2}} + k_1\right) + \left(\frac{k_1}{2}c_2 e^{\frac{k_1}{2}T} + k_1\right)\right\}}{\exp\left\{(k+1)\left(c_3 + c_4 e^{-k_1 T} + e^{\frac{-k_1}{2}T} - k_1 T\right) + c_1 + c_2 e^{\frac{-k_1}{2}T} - k_1 T\right\}}\right](\cosh \lambda) + (\cosh \lambda)'$$

$$\tag{5.2.31}$$

Deceleration parameter q is given by

$$q = -1 - \frac{(k+1)\left(k_1^2 c_4 e^{-k_1 T} + \frac{k_1^2}{4} e^{\frac{-k_1}{2}T}\right) + \left(\frac{k_1^2}{4} c_2 e^{\frac{-k_1}{2}T}\right)}{\left[(k+1)\left(k_1 c_4 e^{-k_1 T} + \frac{k_1}{2} e^{\frac{-k_1}{2}T} + k_1\right) + \left(\frac{k_1}{2} c_2 e^{\frac{k_1}{2}T} + k_1\right)\right]^2} \qquad (5.2.32)$$

A positive sign of the deceleration parameter q indicates the standard decelerating model while the negative sign indicates the inflating model. Recent observations show that the deceleration parameter of the model is in the range of $-1 < q < 0$ and the present day Universe is undergoing an accelerated expansion. From equation (5.2.32), it is observed that the deceleration parameter q is negative for all values of T, hence the obtained model represents an accelerated expansion of the Universe.

The non-vanishing shear scalar components are given by

$$\sigma_{11} = \frac{kk_1^2(\sinh^2\lambda - 1)(\cosh\lambda)\left(c_4 e^{-k_1 T} + \frac{e^{\frac{-k_1}{2}T}}{2} + 1\right)\left\{(k+1)\left(c_4 e^{-k_1 T} + \frac{k_1}{2}e^{\frac{-k_1}{2}T} + k_1\right) + \left(\frac{k_1}{2}c_2 e^{\frac{-k_1}{2}T} + k_1\right) - 1\right\} + (\cosh\lambda)'}{3\exp\left\{(1-k)\left(c_3 + c_4 e^{-k_1 T} + e^{\frac{-k_1}{2}T} - k_1 T\right) + c_1 + c_2 e^{\frac{-k_1}{2}T} - k_1 T\right\}}$$

$$(5.2.33)$$

$$\sigma_{22} = \frac{k_1\left[3\left(c_1 e^{-k_1 T} + \frac{1}{2}e^{\frac{-k_1}{2}T} + 1\right) + (\cosh\lambda)(k+1)\left(c_4 e^{-k_1 T} + \frac{k_1}{2}e^{\frac{-k_1}{2}T} + k_1\right) + \left(\frac{k_1}{2}c_2 e^{\frac{-k_1}{2}T} + k_1\right)\right] - (\cosh\lambda)'}{3\exp\left\{(k-1)\left(c_3 + c_4 e^{-k_1 T} + e^{\frac{-k_1}{2}T} - k_1 T\right) + c_1 + c_2 e^{\frac{-k_1}{2}T} - k_1 T\right\}}$$

$$(5.2.34)$$

$$\sigma_{33} = \frac{k_1\left[\frac{3c_2}{2}e^{\frac{-k_1}{2}T} + 3 + (\cosh\lambda)(k+1)\left(c_4 e^{-k_1 T} + \frac{k_1}{2}e^{\frac{-k_1}{2}T} + k_1\right) + \left(\frac{k_1}{2}c_2 e^{\frac{-k_1}{2}T} + k_1\right)\right] + (\cosh\lambda)'}{3\exp\left\{(k+1)\left(c_3 + c_4 e^{-k_1 T} + e^{\frac{-k_1}{2}T} - k_1 T\right) - c_1 - c_2 e^{\frac{-k_1}{2}T} + k_1 T\right\}}$$

$$(5.2.35)$$

$$\sigma_{44} = \frac{3(\cosh\lambda)' - k_1(1+\cosh^2\lambda)(\cosh\lambda)\left\{(k+1)\left(c_4 e^{-k_1 T} + \frac{k_1}{2}e^{\frac{-k_1}{2}T} + k_1\right) + \left(\frac{k_1}{2}c_2 e^{\frac{-k_1}{2}T} + k_1\right)\right\} + (\cosh\lambda)'}{3\exp\left\{(k+1)\left(c_3 + c_4 e^{-k_1 T} + e^{\frac{-k_1}{2}T} - k_1 T\right) + c_1 + c_2 e^{\frac{-k_1}{2}T} - k_1 T\right\}}$$

$$(5.2.36)$$

$$\sigma_{14} = \frac{3(e^\alpha \sinh\lambda)' + \left\{\left[6(c_4 e^{-k_1 T} + 0.5 e^{\frac{-k_1}{2}T} + 1)\sinh\lambda - (\cosh\lambda)\left[(k+1)\left(c_4 e^{-k_1 T} + \frac{k_1}{2}e^{\frac{-k_1}{2}} + k_1\right) + \left(\frac{k_1}{2}c_2 e^{\frac{k_1}{2}T} + k_1\right)\right]\right]e^\alpha + k_1(\cosh\lambda)'\sinh 2\lambda\right\}}{6\exp\left\{(k+1)\left(c_3 + c_4 e^{-k_1 T} + e^{\frac{-k_1}{2}T} - k_1 T\right) + c_1 + c_2 e^{\frac{-k_1}{2}T} - k_1 T\right\}}$$

$$(5.2.37)$$

The non-vanishing rotational tensor component is given by

$$\omega_{14} = \frac{\left(\exp\left\{ k\left(c_3 + c_4 e^{-k_1 T} + e^{\frac{-k_1}{2}T} - k_1 T \right) \right\} \sinh \lambda \right)'}{\exp\left\{ (k+1)\left(c_3 + c_4 e^{-k_1 T} + e^{\frac{-k_1}{2}T} - k_1 T \right) + c_1 + c_2 e^{\frac{-k_1}{2}T} - k_1 T \right\}}$$

$$(5.2.38)$$

where

$$\sinh \lambda = \frac{1}{2}\left[\frac{(k_4 - k_7)e^{-k_1 T} - (k_3 + k_8)e^{\frac{-k_1 T}{2}} + (k_5 - k_9)e^{\frac{-3k_1 T}{2}} + k_{10}e^{-2k_1 T} - 0.3k_1^2 - k_6}{k_7 e^{-k_1 T} - k_8 e^{\frac{-k_1 T}{2}} + k_9 e^{\frac{-3k_1 T}{2}} - k_{10}e^{-2k_1 T} + 0.3k_1^2} \right]^{\frac{1}{2}}$$

$$\cosh \lambda = \frac{1}{2}\left[\frac{(k_4 + k_7)e^{-k_1 T} + (k_8 - k_3)e^{\frac{-k_1 T}{2}} + (k_5 + k_9)e^{\frac{-3k_1 T}{2}} - k_{10}e^{-2k_1 T} + 0.3k_1^2 - k_6}{k_7 e^{-k_1 T} + k_8 e^{\frac{-k_1 T}{2}} + k_9 e^{\frac{-3k_1 T}{2}} - k_{10}e^{-2k_1 T} + 0.3k_1^2} \right]^{\frac{1}{2}}$$

Conclusions:

Here, we have investigated tilted Bianchi type-*I* cosmological model in Brans-Dicke scalar tensor theory of gravitation. For $k_1 < 0$, the spatial volume doesn't vanish for all values of T and increases exponentially with T. The model is expanding, shearing, rotating and anisotropic. Since deceleration parameter $q < 0$ for all values of T, the obtained model represents the accelerating

Universe. The heat conduction vector $q_1 \to \infty$, at initial epoch whereas $q_1 \to 0$ for large values of T. Also obtained and presented non-vanishing components of shear scalar and rotational vectors.

5.3 Tilted Bianchi Type-*II*, *VIII* and *IX* Models in Lyra Manifold

5.3.1 Metric and field equations:

We consider spatially homogeneous Bianchi type-*II*, *VIII* and *IX* metrics in the form

$$ds^2 = -dt^2 + R^2\left[d\theta^2 + f^2(\theta)d\varphi^2\right] + S^2\left[d\psi + h(\theta)d\varphi\right]^2 \qquad (5.3.1)$$

where θ, φ and ψ are the Eulerian angles. Also R and S are functions of t only.

It represents

Bianchi type-*II*   if $f(\theta) = 1$ and $h(\theta) = \theta$

Bianchi type-*VIII* if $f(\theta) = Cosh\,\theta$ and $h(\theta) = Sinh\theta$

Bianchi type- *IX*   if $f(\theta) = Sin\theta$ and $h(\theta) = Cos\theta$

The field equations given by Sen (1951) are

$$R_{ij} - \frac{1}{2} R g_{ij} + \frac{3}{2} \phi_i \phi^i - \frac{3}{4} g_{ij} \phi_l \phi^l = -8\pi T_{ij} \qquad (5.3.2)$$

where T_{ij} is the energy momentum tensor, ϕ_i is the displacement field and the other symbols have their usual meaning as in Riemannian geometry. The displacement field ϕ_i can be written as $\phi_i = (0,0,0,\beta(t))$.

The energy momentum tensor for a perfect fluid distribution with heat conduction is given by

$$T_j^i = (\rho + p)u_i u^j + p g_i^j + q_i u^j + u_i q^j \qquad (5.3.3)$$

together with

$$g_{ij} u^i u^j = -1, \qquad (5.3.4)$$

$$q_i q^i > 0 \quad \text{and} \quad q_i u^i = 0. \qquad (5.3.5)$$

where p is the pressure and ρ is the energy density of the perfect fluid distribution, q_i is the heat conduction vector orthogonal to u_i.

The fluid vector u_i has the components $u^i = \left(\dfrac{\sinh \lambda}{R}, 0, 0, \cosh \lambda \right)$

satisfying (5.3.5) and λ is the tilt angle.

Using (5.3.3) and (5.3.5) the field equations (5.3.2) for the metric (5.3.1) can be written as

$$\frac{\ddot{R}}{R} + \frac{\ddot{S}}{S} + \frac{\dot{R}\dot{S}}{RS} + \frac{S^2}{4R^4} + \frac{3}{4}\beta^2 = -8\pi \left\{ (\rho + p)\sinh^2 \lambda + p + 2q_1 \left(\frac{\sinh \lambda}{R} \right) \right\}$$

$$(5.3.6)$$

$$2\frac{\ddot{R}}{R} + \frac{\dot{R}^2 + \delta}{R^2} - \frac{3S^2}{4R^4} + \frac{3}{4}\beta^2 = -8\pi p \tag{5.3.7}$$

$$2\frac{\dot{R}\dot{S}}{RS} + \frac{\dot{R}^2 + \delta}{R^2} - \frac{S^2}{4R^4} - \frac{3}{4}\beta^2 = -8\pi \left\{ -(\rho + p)\cosh^2 \lambda + p - 2q_1 \left(\frac{\sinh \lambda}{A} \right) \right\}$$

$$(5.3.8)$$

$$(\rho + p)A\sinh \lambda \cosh \lambda + q_1 \cosh \lambda + q_1 \left(\frac{\sinh^2 \lambda}{\cosh \lambda} \right) = 0 \tag{5.3.9}$$

here the overhead dot denotes differentiation with respect to t.

5.3.2 Solution of the field equations:

The field equations (5.3.6) to (5.3.9) are only four independent equations with seven unknowns R, S, ρ, p, λ, q_1

and β. So, in order to get a deterministic solution we take the following plausible physical conditions:

i. The shear scalar σ is proportional to scalar expansion θ, which leads to the linear relationship between the metric potentials R and S, i.e.,

$$R = S^k \text{ , where } k \neq 1 \tag{5.3.10}$$

ii. Stiff fluid condition i.e.,

$$p = \rho \tag{5.3.11}$$

iii. $\beta = \beta_0 t^n$ (Adhav 2011) $\tag{5.3.12}$

where n and k are arbitrary constants.

From equations (5.3.6), (5.3.8) and (5.3.10), (5.3.11), we get

$$\frac{\ddot{S}}{S} + 2k\frac{\dot{S}^2}{S^2} + \frac{\delta}{(n+1)}S^{-2k} = 0 \tag{5.3.13}$$

Bianchi type-*II* ($\delta = 0$) model:

If $\delta = 0$, from equation (5.3.13), we get

$$\frac{\ddot{S}}{S} + 2k\frac{\dot{S}^2}{S^2} = 0 \tag{5.3.14}$$

From equation (5.3.14), we get

$$S = \left(k_3 t + k_4\right)^{\frac{k_1}{k_3}} \tag{5.3.15}$$

where $k_3 = (2k+1)k_1$, $k_4 = (2k+1)k_2$ and $k_1 \neq 0$, k_2 are integrating constants.

From equations (5.3.10) and (5.3.15), we get

$$R = \left(k_3 t + k_4\right)^{\frac{kk_1}{k_3}} \tag{5.3.16}$$

Now the metric (5.3.1) can be written as

$$ds^2 = -dt^2 + \left(k_3 t + k_4\right)^{\frac{2kk_1}{k_3}} \left(d\theta^2 + d\varphi^2\right) + \left(k_3 t + k_4\right)^{\frac{2k_1}{k_3}} \left(d\psi + \theta\, d\varphi\right)^2 \tag{5.3.17}$$

From equations (5.3.7), (5.3.12), (5.315) and (5.3.16), we get

$$\rho = p = \frac{1}{8\pi}\left[\frac{3}{4}\left(k_3 t + k_4\right)^{\frac{2k_1}{k_3}(1-2k)} + \frac{kk_1(2k_3 - 3kk_1)}{\left(k_3 t + k_4\right)^2} - \frac{3}{4}\beta_0{}^2 t^{2n}\right] \tag{5.3.18}$$

Thus the metric (5.3.17) together with (5.3.18) constitutes a spatially homogeneous and anisotropic Bianchi type-II tilted cosmological model in Sen theory based on Lyra geometry.

From equations (5.3.8), (5.3.9) and (5.3.18), we get

$$q_1 = \frac{1}{8\pi}\left[\frac{kk_1(3kk_1 - 2k_3)}{(k_3t + k_4)^2} - \frac{3}{4}(k_3t + k_4)^{\frac{2k_1}{k_3}(1-2k)} + \frac{3}{4}\beta_0^2 t^{2n}\right] \tag{5.3.19}$$

$$X(k_3t + k_4)^{\frac{kk_1}{k_3}} \tanh 2\lambda \cosh\lambda$$

$$\lambda = \frac{1}{2}\cosh^{-1}\left\{\frac{\dfrac{kk_1(3kk_1 - 2k_3)}{(k_3t + k_4)^2} - \dfrac{3}{4}(k_3t + k_4)^{\frac{2k_1}{k_3}(1-2k)} + \dfrac{3}{4}\beta_0^2 t^{2n}}{\dfrac{k_1^2(k^2 + k + 1) - k_1k_3(k+1)}{(k_3t + k_4)^2} + \dfrac{(k_3t + k_4)^{\frac{2k_1}{k_3}(1-2k)}}{4} + \dfrac{3}{4}\beta_0^2 t^{2n}}\right\} \tag{5.3.20}$$

From figure-5.1, it is observed that the tilt angle doesn't vanish at any time and reaches a constant value for large values time, so the obtained model is tilted throughout the evolution of Universe.

Figure-5.2 demonstrates the behavior of pressure and heat conduction vector versus time. It is observed that both p and q_1

are decreasing functions of time and vanishes for large values of time.

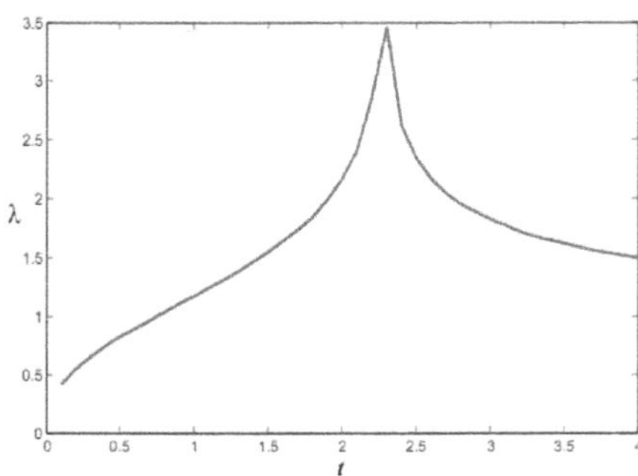

Fig. 5.1: Plot of tilt angle versus time t.

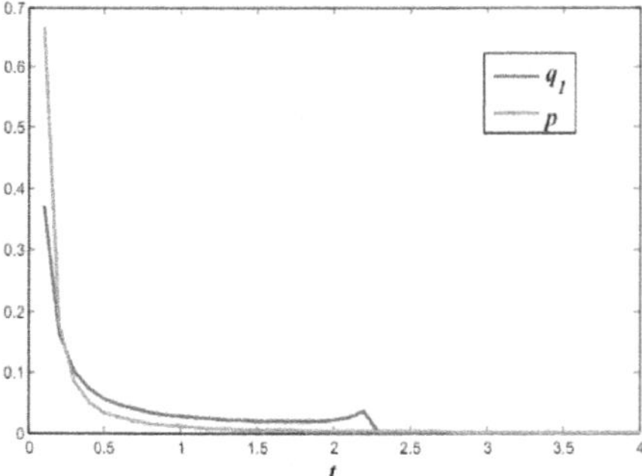

Fig. 5.2: Plot of heat conduction vector and pressure versus t.

Bianchi type-*VIII* ($\delta = -1$) model:

If $\delta = -1$, from equation (5.3.13), we get

$$\frac{\ddot{S}}{S} + 2k\frac{\dot{S}^2}{S^2} - \frac{1}{(k+1)}S^{-2k} = 0 \tag{5.3.21}$$

From equation (5.3.21), we get

$$S = \left(k_6 t + k_7\right)^{1/k} \tag{5.3.22}$$

where $k_6 = \dfrac{k}{(k+1)}$, $k \neq -1$, $k_7 = kk_5$ and k_5 is an integration

constant.

From equations (5.3.10) and (5.3.22), we get

$$R = \left(k_6 t + k_7\right) \tag{5.3.23}$$

Now the metric (5.3.1) can be written as

$$ds^2 = -dt^2 + \left(k_6 t + k_7\right)^2 \left(d\theta^2 + \cosh^2\theta\, d\varphi^2\right) + \left(k_6 t + k_7\right)^{2/k}\left(d\psi + \sinh\theta\, d\varphi\right)^2$$

$$\tag{5.3.24}$$

From equations (5.3.7), (5.3.22) and (5.3.23), we get

the energy density and the pressure as

$$\rho = p = \frac{1}{8\pi}\left[\frac{3}{4}\left(k_6 t + k_7\right)^{\frac{2}{k}-4} + \frac{1-k_6^{\,2}}{\left(k_6 t + k_7\right)^2} - \frac{3}{4}\beta_0^{\,2} t^{2n}\right] \tag{5.3.25}$$

Thus the metric (5.3.24) together with (5.3.25) constitutes a spatially homogeneous and anisotropic Bianchi type-*VIII* tilted cosmological model in Lyra manifold.

From equation (5.3.8), (5.3.9) and (5.3.25), we obtained heat conduction vector and tilt angle as

$$q_1 = \frac{1}{8\pi}\left[\frac{3}{4}\beta_0{}^2 t^{2n} + \frac{k_6{}^2 - 1}{\left(k_6 t + k_7\right)^2} - \frac{3}{4}\left(k_6 t + k_7\right)^{\frac{2}{k}-4}\right]\left(k_6 t + k_7\right)\left(\tanh 2\lambda\right)\left(\cosh\lambda\right)$$

$$(5.3.26)$$

$$\lambda = \frac{1}{2}\cosh^{-1}\left\{\frac{\dfrac{3}{4}\beta_0{}^2 t^{2n} - \dfrac{3}{4}\left(k_6 t + k_7\right)^{\frac{2}{k}-4} + \dfrac{k_6{}^2 - 1}{\left(k_6 t + k_7\right)^2}}{\dfrac{3}{4}\beta_0{}^2 t^{2n} + \dfrac{1}{4}\left(k_6 t + k_7\right)^{\frac{2}{k}-4} + \dfrac{k_6{}^2}{k^2\left(k_6 t + k_7\right)^2}}\right\} \qquad (5.3.27)$$

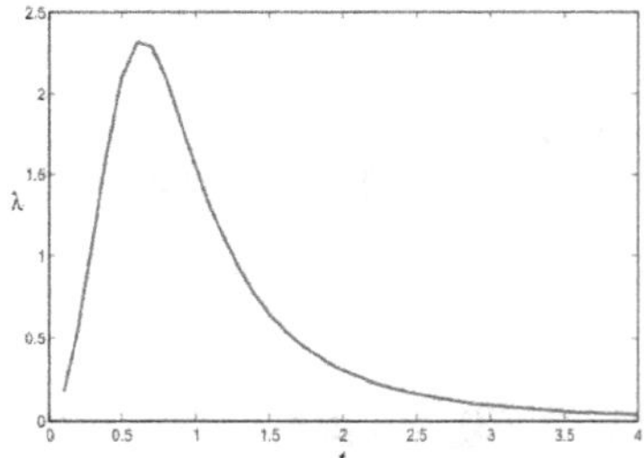

Fig. 5.3: Plot of tilt angle versus t.

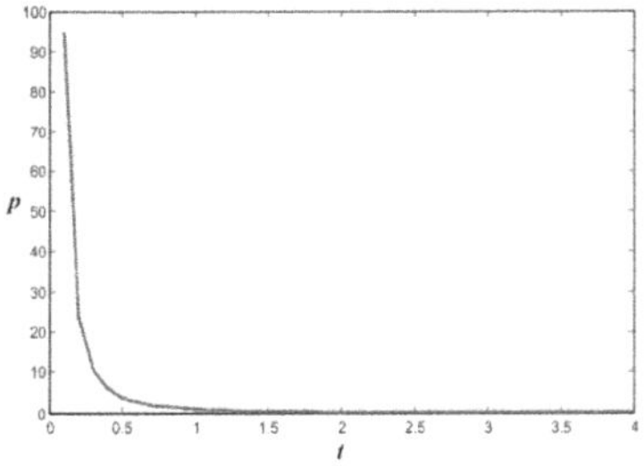

Fig. 5.4: Plot of pressure versus t.

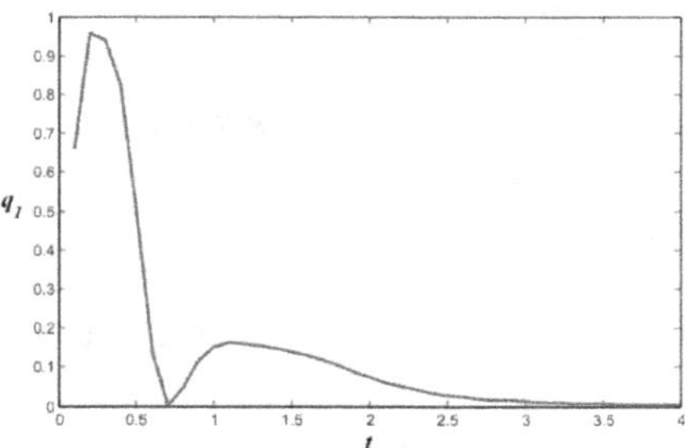

Fig. 5.5: Plot of heat conduction vector versus t.

139

From figure-5.3, we observed that the tilt angle is increased and then decreased   rapidly at initial epoch and finally tends to zero for large values of time. Figure-5.4 explains the behavior of pressure versus time. It is observed that pressure is decreasing function of time and vanishes for large values of time t. From figure-5.5, it is observed that heat conduction vector initially decreases rapidly and vanishes for large values of time t.

Bianchi type-*IX* ($\delta=1$) model:

Here it is observed that the model in this case (i.e. for $\delta=1$) is similar to the model obtained in the case of Bianchi type-*VIII*, where $k_6{}^2 = \dfrac{-1}{(k+1)^2}$. But in this case we get an imaginary solution of the model.

5.3.3 Some important features and conclusions:

Bianchi type-*II* ($\delta=0$) model:

The spatial volume for the model (5.3.17) is

$$V = T^{\frac{k_1}{k_3}(2k+1)},$$

(5.3.28)

where $T = (k_3 t + k_4)$

The expression for expansion scalar θ is given by

$$\theta = \frac{k_1(2k+1)}{T}\left(\frac{2c_4 T^{-2} - T^{c_2} + 3\beta_0^2 t^{2n}}{2c_3 T^{-2} + T^{c_2} + 3\beta_0^2 t^{2n}}\right)^{\frac{1}{2}} + \left(\cosh \lambda\right)' \quad (5.3.29)$$

The deceleration parameter q is given by

$$q = \frac{d}{dt}\left(\frac{1}{H}\right) - 1 = 2 \tag{5.3.30}$$

The non-vanishing shear scalar components are given by

$$\sigma_{11} = T^{\left(\frac{2kk_1}{k_3}-1\right)}\left\{\frac{kk_1}{k_3} - \left(\frac{k_1(2k+1)}{3}(\cosh \lambda) + (\cosh \lambda)'\right)\left(\frac{2(c_1 - c_3)T^{-2} - 6T^{c_2} - 3\beta_0^2 t^{2n}}{2c_3 T^{-2} + 4T^{c_2} + 3\beta_0^2 t^{2n}}\right)\right\}$$

$$\tag{5.3.31}$$

$$\sigma_{22} = \left[k_1\left(\frac{2k+4}{3}\right)(\cosh \lambda) + (\cosh \lambda)'\right]T^{\left(\frac{2k_1}{k_3}-1\right)} \tag{5.3.32}$$

$$\sigma_{44} = \left(\frac{k_1(2k+1)}{T}\cosh \lambda + (\cosh \lambda)'\right)\left(\frac{2(c_4 + c_3)T^{-2} + 3T^{c_2} + 6\beta_0^2 t^{2n}}{6\left(2c_3 T^{-2} + 4T^{c_2} + 3\beta_0^2 t^{2n}\right)}\right) + (\cosh \lambda)'$$

$$\tag{5.3.33}$$

$$\sigma_{14} = \frac{1}{2}T^{\frac{kk_1}{k_3}-1}\left(T(\sinh \lambda)' - kk_1 \sinh \lambda\right) + \frac{1}{6}\left(\frac{k_1(2k+1)}{T}\cosh \lambda + (\cosh \lambda)'\right)T^{\frac{kk_1}{k_3}}\sinh 2\lambda$$

$$\tag{5.3.34}$$

The rotational vector is given by

$$\omega_{14} = T^{\frac{kk_1}{k_3}-1}\left(kk_1 \sinh \lambda + T\left(\sinh \lambda\right)'\right) \tag{5.3.35}$$

where

$$\sinh \lambda = \left(\frac{2\left(c_1 T^{-2} - T^{c_2}\right)}{2c_3 T^{-2} + T^{c_2} + 3\beta_0^2 t^{2n}}\right)^{\frac{1}{2}}, \cosh \lambda = \left(\frac{2c_4 T^{-2} - T^{c_2} + 3\beta_0^2 t^{2n}}{2c_3 T^{-2} + T^{c_2} + 3\beta_0^2 t^{2n}}\right)^{\frac{1}{2}},$$

$$\left(\cosh \lambda\right)' = \frac{\left\{\begin{array}{c}\dfrac{15}{4}\beta_0^2 t^{2n-1}\left[2nT - k_3 c_2 t\right]T^{c_2-1} + 12\left(c_3 - c_4\right)\left(nT + k_3 t\right)T^{-3} \\[2mm] -\dfrac{k_3}{2}\left(c_2 + 2\right)\left(4c_4 + c_3\right)T^{c_2-3}\end{array}\right\}}{\left(c_4 T^{-2} - \dfrac{1}{2}T^{c_2} + \dfrac{3}{2}\beta_0^2 t^{2n}\right)^{\frac{1}{2}}\left(c_3 T^{-2} + 2T^{c_2} + \dfrac{3}{2}\beta_0^2 t^{2n}\right)^{\frac{3}{2}}},$$

$$\left(\sinh \lambda\right)' = \frac{\left\{\begin{array}{c}2c_3 k_3 T^{-5} + \dfrac{3}{2}\beta_0^2 t^{2n}\left[\dfrac{2n}{t}\left(c_2 T - c_1 T^{-2}\right) - c_2 k_3 T^{c_2-1} - 2k_3 T^{-3}\right] \\[2mm] -T^{c_2-3}\left[c_2\left(c_1 + c_3\right) + 2c_3 + 4\right]\end{array}\right\}}{\left(c_1 T^{-2} - T^{c_2}\right)^{\frac{1}{2}}\left(c_3 T^{-2} + 2T^{c_2} + \dfrac{3}{2}\beta_0^2 t^{2n}\right)^{\frac{3}{2}}},$$

$$c_1 = kk_1\left(2kk_1 - k_3\right) - k_1^2\left(k + 1\right) + k_1 k_3, \quad c_2 = \frac{2k_1}{k_3}\left(1 - 2k\right),$$

$$c_3 = 2\left[k_1^2\left(k^2 + 1\right) - k_1 k_3\left(k + 1\right) + kk_1^2\right],$$

$$c_4 = kk_1\left(4kk_1 - 3k_3\right) + k_1^2\left(k + 1\right) - k_1 k_3.$$

Bianchi type-*VIII* ($\delta = -1$) model:

The spatial volume for the models (5.3.24) is

$$V = \tau^{2+\frac{1}{k}} \tag{5.3.36}$$

where $\tau = (k_6 t + k_7)$ and $k_6^2 = \dfrac{1}{(1+k)^2}$.

The expansion scalar θ is given by

$$\theta = \frac{k_6(2k+1)}{k\tau}\left(\frac{2c_8\tau^{-2} - \tau^{c_6} + 3\beta_0^2 t^{2n}}{2c_7\tau^{-2} + \tau^{c_6} + 3\beta_0^2 t^{2n}} \right)^{\frac{1}{2}} + (\cosh \lambda)' \tag{5.3.37}$$

The deceleration parameter q for the model (3.15) is given by

$$q = \frac{d}{dt}\left(\frac{1}{H}\right) - 1 = -1 + \frac{3k}{2k+1} \tag{5.3.38}$$

The non-vanishing shear scalars are given by

$$\sigma_{11} = \tau^{(2k-1)}\left\{ k - \left(\frac{k_6(2k+1)}{3k}(\cosh \lambda) + (\cosh \lambda)' \right)\left(\frac{(c_5 - c_7)\tau^{-2} - 3\tau^{c_6} - \frac{3}{2}\beta_0^2 t^{2n}}{c_7\tau^{-2} + 2\tau^{c_6} + \frac{3}{2}\beta_0^2 t^{2n}} \right) \right\} \tag{5.3.39}$$

$$\sigma_{22} = \left(k_6\left(\frac{2k+4}{3k} \right)(\cosh \lambda) + (\cosh \lambda)' \right)\tau \tag{5.3.40}$$

$$\sigma_{44} = \left(\frac{k_6(2k+1)}{k\tau}(\cosh\lambda) + (\cosh\lambda)' \right) \left(\frac{2(c_8+c_7)\tau^{-2} + 3\tau^{c_6} + 6\beta_0^{\,2}t^{2n}}{3\left(2c_7\tau^{-2} + 4\tau^{c_6} + 3\beta_0^{\,2}t^{2n}\right)} \right) + (\cosh\lambda)'$$

$$(5.3.41)$$

$$\sigma_{14} = \frac{1}{2}\tau^{k-1}\left(\tau(\sinh\lambda)' - k_6\sinh\lambda \right) + \frac{1}{6}\left(\frac{k_6(2k+1)}{k\tau}(\cosh\lambda) + (\cosh\lambda)' \right)\tau^k \sinh 2\lambda$$

$$(5.3.42)$$

The rotational tensor is given by

$$\omega_{14} = \tau^{k-1}\left(k_6\sinh\lambda + \tau(\sinh\lambda)' \right) \qquad\qquad (5.3.43)$$

where $\sinh\lambda = \left(\dfrac{2\left(c_5\tau^{-2} - \tau^{c_6}\right)}{2c_7\tau^{-2} + \tau^{c_6} + 3\beta_0^{\,2}t^{2n}} \right)^{\frac{1}{2}}$,

$$\cosh\lambda = \left(\frac{2c_8\tau^{-2} - \tau^{c_6} + 3\beta_0^{\,2}t^{2n}}{2c_7\tau^{-2} + \tau^{c_6} + 3\beta_0^{\,2}t^{2n}} \right)^{\frac{1}{2}},$$

$$(\cosh\lambda)' = \frac{\left\{ \begin{array}{c} \dfrac{15}{4}\beta_0^{\,2}t^{2n-1}\left[2n\tau - \dfrac{k_6}{k}c_6t \right]\tau^{c_6-1} + 12\left(c_7 - c_8\right)\left(n\tau + k_3t\right)\tau^{-3} \\[4mm] -\dfrac{k_6}{2k}\left(c_6 + 2\right)\left(4c_8 + c_7\right)\tau^{c_6-3} \end{array} \right\}}{\left(c_8\tau^{-2} - \dfrac{1}{2}\tau^{c_2} + \dfrac{3}{2}\beta_0^{\,2}t^{2n} \right)^{\frac{1}{2}}\left(c_7\tau^{-2} + 2\tau^{c_6} + \dfrac{3}{2}\beta_0^{\,2}t^{2n} \right)^{\frac{3}{2}}},$$

$$
(\sinh \lambda)' = \frac{\left\{ \begin{array}{c} \dfrac{2c_7 k_6 \tau^{-5}}{k} + \dfrac{3}{2}\beta_0^{\,2} t^{2n}\left[\dfrac{2n}{t}\left(c_6\tau - c_5\tau^{-2}\right) - \left(\dfrac{c_6 k_6 \tau^{c_6-1} + 2k_6\tau^{-3}}{k}\right)\right] \\[2ex] - \tau^{c_6-3}\left[c_6\left(c_5 + c_7\right) + 2c_7 + 4\right] \end{array}\right\}}{\left(c_5\tau^{-2} - \tau^{c_6}\right)^{\frac{1}{2}}\left(c_7\tau^{-2} + 2\tau^{c_6} + \dfrac{3}{2}\beta_0^{\,2} t^{2n}\right)^{\frac{3}{2}}},
$$

$$
c_5 = k_6^2 - 1 - \frac{k_6^2}{k^2}, \quad c_6 = \frac{2}{k} - 4, \quad c_7 = \frac{2k_6^2}{k^2}, \text{ and } c_8 = k_6^2 - 1 + \frac{k_6^2}{k^2}.
$$

Conclusions:

Here, we have presented Bianchi type-*II,* and *VIII* tilted cosmological models in the frame work of a scalar-tensor theory proposed by Sen (1957) based on Lyra (1951) geometry. The following are some of the conclusions:

- The volume of the model (5.3.17) vanish at $t = \dfrac{-k_4}{k_3}$ and increases as $t \to \infty$. The model has no singularity for $k > 0$. Also, we observe that, the expansion scalar is decreasing function of time.

- The deceleration parameter q is positive for the model (5.3.17). So, model represents standard decelerating Universe. Viswakarma (2003) has shown that decelerating models are also consistent with recent cosmic background observations made by WAMP. However, in spite of the fact that the Universe, in this case, decelerates in the standard way it will accelerate in finite time due to cosmic re collapse where the Universe in turns inflates "decelerates and then accelerates" (Nojiri and Odintsov, 2003b).

- The volume of the model (5.3.24) vanish at $t = \dfrac{-k_7}{k_6}$ and increases as $t \to \infty$. The model is singularity free for $k > 0$.

- For model (5.3.24), the deceleration parameter $q < 0$ for $k < 1$. This shows that the model represents accelerated expansion of the Universe.

- We find that the pressure, energy density, heat conduction vector of the fluid distribution and tilt angle are decreasing functions of time and vanishes for large values of time.

Chapter-6

Kantowski-Sachs two fluid radiating models in Brans-Dicke theory of gravitation*

***The work presented in this chapter is communicated to following journal:**

➢ **Kantowski-Sachs two fluid radiating models in Brans-Dicke theory of gravitation: Journal of Advances in Astrophysics.**

6.1 Introduction

In recent years there has been huge interest in cosmological models with dark energy in general relativity because of the fact that our observable Universe is undergoing a phase of accelerated expansion which has been confirmed by several cosmological observations such as type *I*a supernova by several authors (Reiss et al. 1998, 2000, 2004; Perlmutter et al. 1997, 1998, 1999, 2003). Caldwell (2002) and Huange (2006) have discussed cosmic microwave background anisotropy and Daniel et al. (2008) have studied large scale structure and strongly indicate that dark energy dominates the present Universe, causing cosmic acceleration. Based on these observations, cosmologists have accepted the idea of dark energy, which is a fluid with negative pressure making up around 70 % of the present Universe energy content to be responsible for this acceleration due to repulsive gravitation. We have discussed in detail the concept of dark energy in chapter-1. However, we make a brief mention of the same for on the spot reference.

The simplest candidate of dark energy is the cosmological constant. It is, however, plagued with the so-called coincidence problem and the cosmological constant problem (Weinberg 1989; Sahni and Starobinsky 2000; Peebles and Ratra 2003; Padmanabhan 2003; Copeland et al. 2006). Another possible form of dark energy is provided by scalar fields. Thus some dynamical scalar field, such as quintessence (Wetterich 1988; Ratra and Peebles 1988; Caldwell et al. 1998), phantom (Caldwell 2002; Nojiri and Odintsov 2003a; Nojiri et al. 2005; Sahni and Shtanov 2003; Xiangyun et al. 2008), quintom (Elizalde et al. 2004; Feng et al. 2005; Setare 2006; Cai et al. 2007) and K-essence (Alimohammadi and Sadjadi 2006; Wei et al. 2007), are proposed as possible candidate of dark energy. However, it is worth mentioned here that for these scalar field models the coincident problem still remains. Although the two dark components are usually studied under the assumption that there is no interaction between them, one cannot exclude such a possibility. In fact, researches show that a presumed interaction may help alleviate the coincident problem (Chimento et al. 2003; Chimento and Pavon 2006).

Adhav et al. (2011) have investigated interacting cosmic fluids in LRS Bianchi type-I cosmological model. Saha et al. (2012) revisited two-fluid scenario for dark energy models in an FRW Universe investigated by Amirhashchi et al. (2011). Adhav et al. (2012) have studied Kaluza-Klein interacting cosmic fluid cosmological model. Reddy and Santhi Kumar (2013) have discussed two fluid scenario for dark energy model in a scalar-tensor theory of gravitation. Amirhashchi et al. (2013) have studied interacting two-fluid viscous dark energy models in a non-flat Universe.

Recently, alternative theories of gravitation are attracting more and more attention of researchers. In particular scalar-tensor theories of gravitation proposed by Brans-Dicke (1961) and Saez-Ballester (1986) are playing vital role in the discussion of modern cosmology. The latest inflationary models (Mathiazhagan and Johri 1984), extended inflation (La et al.1990; Steinhardt and Accetta 1990) hyper extended inflation and extended chaotic inflation (Linde 1990) are based on Brans-Dicke theory of gravitation. A detailed discussion of Brans-Dicke theory is presented in chapter-1 and the work done in this theory is presented in chapter-5.

The importance of Kantowski-Sachs space-time is presented in chapter-4. Samanta (2013), Reddy et al. (2014a, b), Pawar and Solanke (2014), Rodrigue et al. (2015), DasuNaidu et al. (2015a, b) are some of the authors who have investigated several aspects of Kantowski-Sachs cosmological models in various scalar-tensor theories of gravitation. Rao and Suryanarayana (2015b) have investigated Kantowski-Sachs cosmological model in $f(R,T)$ theory of gravity.

Motivated by above investigations and discussion, in this chapter, we obtain, spatially homogeneous Kantowski-Sachs cosmological models filled with barotropic fluid and dark energy in Brans-Dicke (1961) theory of gravitation.

The format of the chapter is as follows: we derived the field equations of Brans-Dicke with the help of Kantowski-Sachs metric in section 6.2. Solutions of the field equations in non-interacting and interacting cases are obtained in section 6.3. In section 6.4 we discussed some important features of the obtained models. The last section 6.5 contains some conclusions.

6.2 Metric and Field Equations

We consider a spatially homogeneous Kantowski-Sachs metric of the form

$$ds^2 = dt^2 - A^2 dr^2 - B^2(d\theta^2 + Sin^2\theta \, d\varphi^2) \qquad (6.2.1)$$

where A and B are the functions of time t only.

Brans-Dicke (1961) theory of gravitation is a natural extension of general relativity which introduces an additional scalar field ϕ besides the metric tensor g_{ij} and dimensionless coupling constant ω. The Brans-Dicke field equations for combined scalar and tensor field are given by

$$R_{ij} - \frac{1}{2}Rg_{ij} = -8\pi\phi^{-1}T_{ij} - \omega\phi^{-2}\left(\phi_{,i}\phi_{,j} - \frac{1}{2}g_{ij}\phi_{,r}\phi^{,r}\right) - \phi^{-1}(\phi_{i;j} - g_{ij}\phi_{,r}^{,r}) \qquad (6.2.2)$$

and $\qquad \phi_{,r}^{,r} = 8\pi(3+2\omega)^{-1}T \qquad (6.2.3)$

where R is the scalar curvature, ω and r are constants, T_{ij} is the stress energy tensor of the matter and comma and semicolon denote partial and covariant differentiation respectively.

Also, we have energy conservation equation

$$T^{ij}_{;j} = 0 \tag{6.2.4}$$

This equation is a consequence of the field equations (6.2.2) and (6.2.3).

The total energy momentum tensor for two fluids is given by

$$T_{ij} = (\rho_{tot} + p_{tot})u_i u_j - p_{tot}\, g_{ij} \tag{6.2.5}$$

where $\rho_{tot} = \rho_m + \rho_D$ and $p_{tot} = p_m + p_D$. Here ρ_m and p_m are energy density and pressure of barotropic fluid and ρ_D and p_D are energy density and pressure of dark fluid respectively, u^i is the four-velocity of the fluid satisfying the following condition,

$$g_{ij}u^i u^j = 1 \tag{6.2.6}$$

In a commoving coordinate system, we get

$$T_1^1 = T_2^2 = T_3^3 = -p_{tot}, \quad T_4^4 = \rho_{tot} \quad \text{and} \quad T_j^i = 0 \text{ for } i \neq j \tag{6.2.7}$$

where the quantities ρ_{tot} and p_{tot} are functions of 't' only.

6.3 Solution of the Field Equations

Now with the help of (6.2.7), the field equations (6.2.2) and (6.2.3) for the metric (6.2.1) can be written as

$$2\frac{\ddot{B}}{B}+\frac{\dot{B}^2}{B^2}+\frac{1}{B^2}+\frac{\omega\dot{\phi}^2}{2\phi^2}+\frac{\ddot{\phi}}{\phi}+2\frac{\dot{\phi}\dot{B}}{\phi B}=-\frac{8\pi\,p_{tot}}{\phi} \tag{6.3.1}$$

$$\frac{\ddot{A}}{A}+\frac{\ddot{B}}{B}+\frac{\dot{A}\dot{B}}{AB}+\frac{\omega\dot{\phi}^2}{2\phi^2}+\frac{\ddot{\phi}}{\phi}+\frac{\dot{\phi}}{\phi}\left(\frac{\dot{A}}{A}+\frac{\dot{B}}{B}\right)=-\frac{8\pi\,p_{tot}}{\phi} \tag{6.3.2}$$

$$2\frac{\dot{A}\dot{B}}{AB}+\frac{\dot{B}^2}{B^2}+\frac{1}{B^2}-\frac{\omega\dot{\phi}^2}{2\phi^2}+\frac{\dot{\phi}}{\phi}\left(\frac{\dot{A}}{A}+2\frac{\dot{B}}{B}\right)=\frac{8\pi\,\rho_{tot}}{\phi} \tag{6.3.3}$$

$$\ddot{\phi}+\dot{\phi}\left(\frac{\dot{A}}{A}+2\frac{\dot{B}}{B}\right)=\frac{8\pi}{(3+2\omega)}\left(\rho_{tot}-3p_{tot}\right). \tag{6.3.4}$$

Also the energy conservation equation leads to

$$\dot{\rho}_{tot}+3H\left(\rho_{tot}+p_{tot}\right)=0\,. \tag{6.3.5}$$

Here the over head dot denotes differentiation with respect to t.

The equations (6.3.1) to (6.3.4) is a system of four independent equations with five unknowns $A, B, p_{tot}, \rho_{tot}$ and ϕ. In order to get a deterministic solution we take the following plausible physical condition, the shear scalar σ is proportional to scalar expansion θ, which leads to the linear relationship between the metric potentials A and B,

$$\text{i.e., } B = A^n, \text{ where } n > 1 \text{ is a constant.} \tag{6.3.6}$$

Using (6.3.6) the above field equations (6.3.1) to (6.3.4) can be written as

$$2\frac{\ddot{B}}{B} + \frac{\dot{B}^2}{B^2} + \frac{1}{B^2} + \frac{\omega\dot{\phi}^2}{2\phi^2} + \frac{\ddot{\phi}}{\phi} + 2\frac{\dot{B}\dot{\phi}}{B\phi} = -\frac{8\pi\, p_{tot}}{\phi} \tag{6.3.7}$$

$$\left(n+1\right)\frac{\ddot{B}}{B} + n^2\frac{\dot{B}^2}{B^2} + \frac{\omega\dot{\phi}^2}{2\phi^2} + \frac{\ddot{\phi}}{\phi} + \left(n+1\right)\frac{\dot{B}\dot{\phi}}{B\phi} = -\frac{8\pi\, p_{tot}}{\phi} \tag{6.3.8}$$

$$\left(2n+1\right)\frac{\dot{B}^2}{B^2} + \frac{1}{B^2} - \frac{\omega\dot{\phi}^2}{2\phi^2} + \left(n+2\right)\frac{\dot{B}\dot{\phi}}{B\phi} = \frac{8\pi\, p_{tot}}{\phi} \tag{6.3.9}$$

$$\ddot{\phi} + \left(n+2\right)\frac{\dot{B}\dot{\phi}}{B\phi} = \frac{8\pi}{(3+2\omega)}\left(\rho_{tot} - 3p_{tot}\right) \tag{6.3.10}$$

By considering the radiating model

$$\text{i.e. } \rho_{tot} = 3p_{tot}, \tag{6.3.11}$$

the field equation (6.3.10) can be written as

$$\ddot{\phi} + \left(n+2\right)\frac{\dot{B}\dot{\phi}}{B\phi} = 0 \tag{6.3.12}$$

From equations (6.3.7), (6.3.8) and (6.3.12), we get

$$\frac{\ddot{B}}{B} + (n+1)\frac{\dot{B}^2}{B^2} - \frac{1}{B^2(n-1)} = \frac{\ddot{\phi}}{\phi(n+2)} \tag{6.3.13}$$

From equation (6.3.13), we get

$$A = \left(\frac{k_1}{k_2}Sin(k_2 t + k_3)\right)^n \tag{6.3.14}$$

$$B = \left(\frac{k_1}{k_2}\right)Sin(k_2 t + k_3) \tag{6.3.15}$$

where $k_2^2 = \frac{m^2}{n+2}$, $k_1^2 = \frac{1}{n^2-1}$, $(n\neq 1)$, m is an arbitrary constant

and k_3 is an integrating constant.

Also, the scalar field is given by

$$\phi = k_4 Cos k_6 t + k_5 Sin k_6 t \tag{6.3.16}$$

where k_4, k_5 are integrating constants and $k_6 = m\sqrt{n+2}$.

Thus the metric (6.2.1) can be written as

$$ds^2 = dt^2 - \left(\frac{k_1}{k_2}Sin(k_2 t + k_3)\right)^{2n} dr^2 - \left(\frac{k_1}{k_2}Sin(k_2 t + k_3)\right)^2 (d\theta^2 + Sin^2\theta\, d\varphi^2)$$

$$\tag{6.3.17}$$

The equation of state parameter of the barotropic fluid and dark energy are respectively given by

$$w_m = \frac{p_m}{\rho_m} \text{ and } w_D = \frac{p_D}{\rho_D} . \tag{6.3.18}$$

In the following sub sections we deal the two cases:

(i) non-interacting two-fluid model

(ii) interacting two-fluid model.

6.3.1 Non-interacting two-fluid model:

First, we consider that two fluids do not interact with each other. Therefore, the general form of conservation equation (6.3.5) leads us to write the conservation equation for the dark and barotropic fluid separately as,

$$\dot{\rho}_m + 3H(\rho_m + p_m) = 0 \tag{6.3.19}$$

and $\quad \dot{\rho}_D + 3H(\rho_D + p_D) = 0. \tag{6.3.20}$

Here is, of course, a structural difference between equations (6.3.19) and (6.3.20). Because equation (6.3.19) is in the form of w_m which is constant and hence equation (6.3.19) is integrable. But equation (6.3.20) is a function of w_D. Accordingly, ρ_D and p_D are also function of w_D. Therefore, we cannot integrate equation (6.3.20) as it is a function of w_D which is an unknown time dependent parameter.

By integrating equation (6.3.19) leads to

$$8\pi\rho_m = 8\pi\rho_0 \left[Sin\left(k_2 t + k_3\right)\right]^{-\left(n+2\right)\left(1+w_m\right)}$$

(6.3.21)

where ρ_0 is an integrating constant.

From equations (6.3.9), (6.3.14) - (6.3.16) and (6.3.21), we get

$$8\pi\rho_D = k_2^2 \left\{ \left(1 + 2n + \frac{1}{k_1^2}\right)Cot^2\left(k_2 t + k_3\right) + \frac{1}{k_1^2}\right\}\left(k_4 Cos k_6 t + k_5 Sin k_6 t\right) +$$

$$k_6\left\{ (n+2)k_2 Cot\left(k_2 t + k_3\right) - \frac{\omega}{2}\left(\frac{-k_4 Sin k_6 t + k_5 Cos k_6 t}{\left(k_4 Cos k_6 t + k_5 Sin k_6 t\right)}\right)\right\}\left(-k_4 Sin k_6 t + k_5 Cos k_6 t\right)$$

$$-8\pi\rho_0 \left[Sin\left(k_2 t + k_3\right)\right]^{-\left(n+2\right)\left(1+w_m\right)}$$

by taking $k_4 = Cos k_7$, $k_5 = Sin k_7$, the above equation reduces to

$$8\pi\rho_D = k_2^2 \left\{ \left(1 + 2n + \frac{1}{k_1^2}\right)Cot^2\left(k_2 t + k_3\right) + \frac{1}{k_1^2}\right\}Cos\left(k_6 t - k_7\right)$$

$$- k_6\left\{ (n+2)k_2 Cot\left(k_2 t + k_3\right) + \frac{\omega}{2}k_6 Tan\left(k_6 t - k_7\right)\right\}Sin\left(k_6 t - k_7\right)$$

$$- 8\pi\rho_0 \left[Sin\left(k_2 t + k_3\right)\right]^{-\left(n+2\right)\left(1+w_m\right)}$$

(6.3.22)

From equations (6.3.18) and (6.3.21), we get

$$8\pi\, p_m = 8\pi\rho_0 w_m \left[Sin\left(k_2 t + k_3\right)\right]^{-(n+2)(1+w_m)}$$

$$(6.3.23)$$

From equations (6.3.8), (6.3.14)-(6.3.16) and (6.3.23), we get

$$8\pi\, p_D = \left\{ k_6^2 - k_2^2 \left(n^2 Cot^2\left(k_2 t + k_3\right) - n - 1\right)\right\}\left(k_4 Cos k_6 t + k_5 Sin k_6 t\right) +$$

$$k_6\left\{ (n+1)k_2 Cot(k_2 t + k_3) + \frac{\omega}{2}k_6\left(\frac{-k_4 Sin k_6 t + k_5 Cos k_6 t}{\left(k_4 Cos k_6 t + k_5 Sin k_6 t\right)}\right)\right\}\left(k_4 Sin k_6 t + k_5 Cos k_6 t\right)$$

$$-8\pi\rho_0\left[Sin(k_2 t + k_3)\right]^{-(n+2)(1+w_m)}$$

by taking $k_4 = Cos k_7$, $k_5 = Sin k_7$, the above equation reduces to

$$8\pi\, p_D = \left\{ k_6^2 - k_2^2\left(n^2 Cot^2\left(k_2 t + k_3\right) - n - 1\right)\right\} Cos\left(k_6 t - k_7\right)$$

$$+ k_6\left\{ (n+1)k_2 Cot\left(k_2 t + k_3\right) - \frac{\omega}{2}k_6 Tan\left(k_6 t - k_7\right)\right\} Sin\left(k_6 t - k_7\right)$$

$$- 8\pi\rho_0 w_m\left[Sin\left(k_2 t + k_3\right)\right]^{-(n+2)(1+w_m)}$$

$$(6.3.24)$$

From equations (6.3.18), (6.3.22) and (6.3.24), we get

$$w_D = \frac{\begin{aligned}&\left\{ k_6^2 - k_2^2\left(n^2 Cot^2\left(k_2 t + k_3\right) - n - 1\right)\right\} Cos\left(k_6 t - k_7\right) - 8\pi\rho_0 w_m\left[Sin\left(k_2 t + k_3\right)\right]^{-(n+2)(1+w_m)}\\ &\quad + k_6\left\{ (n+1)k_2 Cot\left(k_2 t + k_3\right) - \frac{\omega}{2}k_6 Tan\left(k_6 t - k_7\right)\right\} Sin\left(k_6 t - k_7\right)\end{aligned}}{\begin{aligned}&k_2^2\left\{\left(1 + 2n + \frac{1}{k_1^2}\right)Cot^2\left(k_2 t + k_3\right) + \frac{1}{k_1^2}\right\} Cos\left(k_6 t - k_7\right) - 8\pi\rho_0\left[Sin\left(k_2 t + k_3\right)\right]^{-(n+2)(1+w_m)}\\ &\quad - k_6\left\{ (n+2)k_2 Cot\left(k_2 t + k_3\right) + \frac{\omega}{2}k_6 Tan\left(k_6 t - k_7\right)\right\} Sin\left(k_6 t - k_7\right)\end{aligned}}$$

$$(6.3.25)$$

Thus the metric (6.3.17) together with (6.3.16) and (6.3.21)-(6.3.25) constitutes Kantowski-Sachs non-interacting two fluid cosmological model in Brans-Dicke scalar-tensor theory of gravitation.

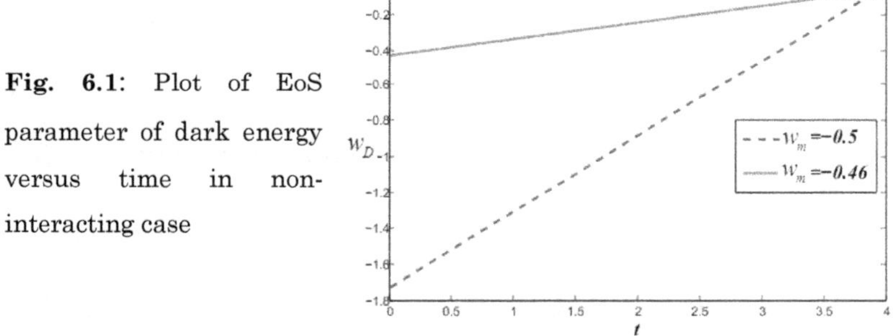

Fig. 6.1: Plot of EoS parameter of dark energy versus time in non-interacting case

Figure-6.1 describes the behavior of EoS parameter of dark energy versus time. It is observed that EoS parameter is increasing function of time and varying in phantom region $(-3 < w_D < -1)$ and quintessence region $(w_D > -1)$. The rapid increase in early stage of Universe depends on EoS parameter of ordinary matter w_m. Also, it is in good agreement with the theoretical results as well as the recent observations.

6.3.2 Interacting two-fluid model:

Secondly, we consider the interaction between dark energy and barotropic fluids. For this purpose we can write the continuity equations for dark fluid and barotropic fluid as

$$\dot{\rho}_m + 3H(\rho_m + p_m) = Q \qquad (6.3.26)$$

and $\quad \dot{\rho}_D + 3H(\rho_D + p_D) = -Q \qquad (6.3.27)$

The quantity Q expresses the interaction between the dark energy components. Since we are interested in an energy transfer from the dark energy to dark matter, we consider $Q > 0$ which ensures that the second law of thermodynamics is fulfilled (Pavon and Wang 2009). Here we emphasize that the continuity equations (6.3.26) and (6.3.27) imply that the interaction term (Q) should be proportional to a quantity with units of inverse of time i.e. $Q \propto 1/t$. Therefore, a first and natural candidate can be the Hubble factor H multiplied with the energy density. Following Amendola et al. (2007) and Guo et al. (2007), we consider

$$Q = 3H\sigma\rho_m \qquad (6.3.28)$$

where σ is a coupling constant.

Using equation (6.3.28) in equation (6.3.26) and after integrating, we obtain

$$8\pi\rho_m = 8\pi\rho_0\left[\text{Sin}\left(k_2 t + k_3\right)\right]^{-\left(n+2\right)\left(1+w_m-\sigma\right)}$$

(6.3.29)

From equations (6.3.9), (6.3.14)-(6.3.16) and (6.3.29), we get energy density for dark energy as

$$8\pi\rho_D = k_2^2\left\{\left(1+2n+\frac{1}{k_1^2}\right)Cot^2\left(k_2 t + k_3\right)+\frac{1}{k_1^2}\right\}\left(k_4 Cos k_6 t + k_5 Sin k_6 t\right)+$$

$$k_6\left\{\left(n+2\right)k_2 Cot\left(k_2 t + k_3\right)-\frac{\omega}{2}\left(\frac{-k_4 Sin k_6 t + k_5 Cos k_6 t}{\left(k_4 Cos k_6 t + k_5 Sin k_6 t\right)}\right)\right\}\left(-k_4 Sin k_6 t + k_5 Cos k_6 t\right)$$

$$-8\pi\rho_0\left[Sin\left(k_2 t + k_3\right)\right]^{\left(n+2\right)\left(1+w_m-\sigma\right)}$$

by taking $k_4 = Cos k_7$, $k_5 = Sin k_7$, the above equation reduces to

$$8\pi\rho_D = k_2^2\left\{\left(1+2n+\frac{1}{k_1^2}\right)Cot^2\left(k_2 t + k_3\right)+\frac{1}{k_1^2}\right\}Cos\left(k_6 t - k_7\right)-$$

$$k_6\left\{\left(n+2\right)k_2 Cot\left(k_2 t + k_3\right)+\frac{\omega}{2}Tan\left(k_6 t + k_7\right)\right\}Sin\left(k_6 t - k_7\right)$$

$$-8\pi\rho_0\left[Sin\left(k_2 t + k_3\right)\right]^{-\left(n+2\right)\left(1+w_m-\sigma\right)}$$

(6.3.30)

From equations (6.3.18) and (6.3.29), we get

$$8\pi p_m = 8\pi w_m \rho_0\left[\text{Sin}\left(k_2 t + k_3\right)\right]^{-\left(n+2\right)\left(1+w_m-\sigma\right)}$$

(6.3.31)

From equations (6.3.7), (6.3.14) - (6.3.16) and (6.3.31) we get dark energy pressure as

$$8\pi\, p_D = \left\{k_6^2 - k_2^2\left(n^2 Cot^2(k_2 t + k_3) - n - 1\right)\right\}(k_4 Cos k_6 t + k_5 Sin k_6 t) +$$

$$k_6\left\{(n+1)k_2 Cot(k_2 t + k_3) + \frac{\omega}{2}k_6\left(\frac{-k_4 Sin k_6 t + k_5 Cos k_6 t}{(k_4 Cos k_6 t + k_5 Sin k_6 t)}\right)\right\}(k_4 Sin k_6 t + k_5 Cos k_6 t)$$

$$-8\pi\rho_0\left[Sin(k_2 t + k_3)\right]^{-(n+2)(1+w_m-\sigma)}$$

by taking $k_4 = Cos k_7$, $k_5 = Sin k_7$, the above equation reduces to

$$8\pi\, p_D = \left\{k_6^2 - k_2^2\left(n^2 Cot^2(k_2 t + k_3) - n - 1\right)\right\}(k_4 Cos k_6 t + k_5 Sin k_6 t) +$$

$$k_6\left\{(n+1)k_2 Cot(k_2 t + k_3) - \frac{\omega}{2}k_6 Tan(k_6 t - k_7)\right\}Sin(k_6 t - k_7) \tag{6.3.32}$$

$$-8\pi\rho_0\left[Sin(k_2 t + k_3)\right]^{-(n+2)(1+w_m-\sigma)}$$

From equations (6.3.18), (6.3.30) and (6.3.32), we get EoS

parameter for dark energy as

$$w_D = \frac{\begin{aligned}&\pi\, p_D = \left\{k_6^2 - k_2^2\left(n^2 Cot^2(k_2 t + k_3) - n - 1\right)\right\}(k_4 Cos k_6 t + k_5 Sin k_6 t) +\\&k_6\left\{(n+1)k_2 Cot(k_2 t + k_3) - \frac{\omega}{2}k_6 Tan(k_6 t - k_7)\right\}Sin(k_6 t - k_7)\\&-8\pi\rho_0\left[Sin(k_2 t + k_3)\right]^{-(n+2)(1+w_m-\sigma)}\end{aligned}}{\begin{aligned}&k_2^2\left\{\left(1+2n+\frac{1}{k_1^2}\right)Cot^2(k_2 t + k_3) + \frac{1}{k_1^2}\right\}Cos(k_6 t - k_7) -\\&k_6\left\{(n+2)k_2 Cot(k_2 t + k_3) + \frac{\omega}{2}Tan(k_6 t + k_7)\right\}Sin(k_6 t - k_7)\\&-8\pi\rho_0\left[Sin(k_2 t + k_3)\right]^{-(n+2)(1+w_m-\sigma)}\end{aligned}} \tag{6.3.33}$$

Thus the metric (6.3.17) together with (6.3.16) and (6.3.29)-

(6.3.33) constitutes Kantowski-Sachs interacting two fluid

cosmological model in Brans-Dicke scalar-tensor theory of gravitation.

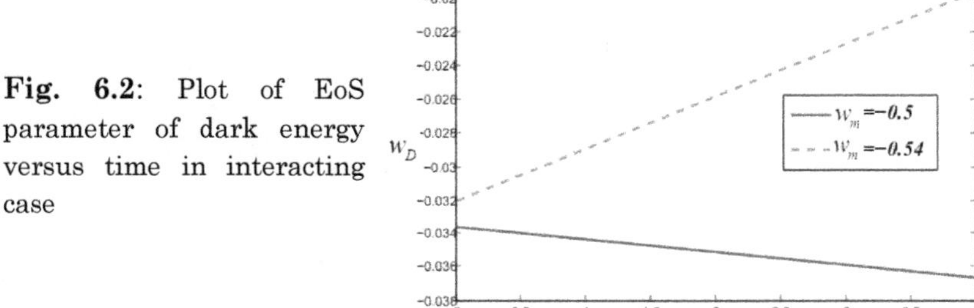

Fig. 6.2: Plot of EoS parameter of dark energy versus time in interacting case

Figure-6.2 demonstrates the behavior of EoS parameter of dark energy versus time. It is observed that EoS parameter is varying in quintessence region only.

6.4 Some Important Features of the Models

The spatial volume and average scale factor for the models are given by

$$V = (-g)^{\frac{1}{2}} = \left(\frac{k_1}{k_2} Sin\left(k_2 t + k_3\right) \right)^{n+2} Sin\theta \qquad (6.4.1)$$

$$a(t) = V^{\frac{1}{3}} = \left(\frac{k_1}{k_2} Sin\left(k_2 t + k_3\right) \right)^{\frac{n+2}{3}} Sin^{\frac{1}{3}}\theta \qquad (6.4.2)$$

163

The expansion scalar θ is given by

$$\theta = u^i_{\ ,i} = k_2(n+2)Cot(k_2 t + k_3) \tag{6.4.3}$$

The shear scalar is given by

$$\sigma^2 = \frac{1}{2}\sigma^{ij}\sigma_{ij} = \frac{1}{2}\left(k_2(n+2)Cot(k_2 t + k_3)\right)^2 \tag{6.4.4}$$

The deceleration parameter q is given by

$$q = (-3\theta^{-2})(\theta_{,i}u^i + \frac{1}{3}\theta^2) = -1 + \frac{3}{n+2}Sec^2(k_2 t + k_3) \tag{6.4.5}$$

The deceleration parameter appears with negative sign implies accelerating expansion of the Universe, which is consistent with the present day observations.

The Hubble's parameter H is given by

$$H = \frac{1}{3}k_2(n+2)Cot(k_2 t + k_3) \tag{6.4.6}$$

The mean anisotropy parameter A_m is given by

$$A_m = 2\left(\frac{n-1}{n+2}\right)^2 \tag{6.4.7}$$

The tensor of rotation $w_{ij} = u_{i,j} - u_{j,i}$ is identically zero and hence this Universe is non- rotational.

6.5 Conclusions

In this chapter, we have obtained and presented spatially homogeneous Kantowski-Sachs cosmological models filled with barotropic fluid and dark energy in a scalar-tensor theory of gravitation proposed by Brans and Dicke (1961). The following are the observations and conclusions:

The obtained models have singularity at $t = t^*$, where $t^* = -k_3 / k_2$. The volume of the models vanishes at $t = t^*$ and expansion scalar is tends to infinity, which shows that the Universe starts evolving with zero volume at $t = t^*$ with an infinite rate of expansion.

From equation (6.4.5), it is observed that deceleration parameter is negative for small values of t and hence the obtained models represent accelerated expansion of the Universe at initial epoch. From equation (6.4.7), one can observe that $A_m \neq 0$ and this indicates that these models are anisotropic.

The two models presented here are anisotropic, non-rotating, shearing and also accelerating.

Chapter-7

Kantowski-Sachs holographic model in Saez-Ballester theory and Kaluza–Klein holographic model in Brans-Dicke theory*

***The work presented in this chapter is communicated to the following journals:**

➢ **Kantowski-Sachs holographic cosmological model in Saez-Ballester theory of gravitation: International Journal of Theoretical Physics.**

➢ **Kaluza-Klein holographic cosmological models in Brans-Dicke theory of gravitation: Turkish Journal of Physics.**

7.1 Introduction

In recent years there has been an immense interest in the study of higher dimensional space-time because of the fact that the cosmos at its early stage of evolution of the universe might have had a higher dimensional era. It is well known that solutions of field equations in general relativity and in scalar-tensor theories in higher dimensional space-time are of physical relevance possibly at the early times before the universe has undergone compactification transitions. Also Marciano (1984) has suggested that the experimental observation if the fundamental constants with varying time could produce the evidence of extra dimensions. A lucid presentation of higher dimensions, in particular Kaluza-Klein space-time in chapter-1 and the work done in Kaluza-Klein space-time is presented in chapter-4. Sharif and Khanum (2011) have studied Kaluza-Klein cosmology with varying G. Sharif and Jawad (2012) have discussed interacting modified holographic dark energy in Kaluza-Klein Universe. Reddy et al. (2012b) have obtained Kaluza-Klein cosmological model in $f(R,T)$ gravity. Samanta and Debata (2013) have

discussed two-fluid cosmological models in Kaluza-Klein space
time. Reddy et al. (2013b) have studied Kaluza-Klein universe
with cosmic strings and bulk viscosity in a scalar-tensor theory of
gravitation. Shri Ram and Priyanka (2013) have discussed some
Kaluza-Klein cosmological models in $f(R,T)$ gravity theory. Reddy
and Vijaya Lakshmi (2015) have studied Kaluza-Klein dark
energy model in Brans-Dicke theory of gravitation.
Mukhopadhyay et al. (2015) have obtained a dark energy model in
Kaluza-Klein cosmology. Rao et al. (2015e) have discussed five
dimensional FRW cosmological models in a scalar-tensor theory of
gravitation.

The significance of Kantowski-Sachs space-time and the
work done in it is, in detail, discussed in chapters-4 and 5. Reddy
et al. (2012b) have obtained five dimensional dark energy model in
a scalar-tensor theory of gravitation. Reddy et al. (2014b) have
discussed Kantowski–Sachs bulk viscous cosmological model in a
scalar–tensor theory of gravitation. Sing and Agrawal (1991) have
discussed Kantowski-Sachs type models in Saez and Ballester
scalar-tensor theory. Mete et al. (2015) have obtained axially

symmetric cosmological model with bulk stress in Saez-Ballester theory of gravitation. Rao and Jayasudha (2015b) have studied Five dimensional spherically symmetric perfect fluid cosmological model in a scalar-tensor theory of gravitation.

We have discussed in detail the concepts of dark energy and holographic dark energy in chapter-1. However, we make brief mention of the same for on the spot reference.

The cosmological observations of type Ia Supernovae (SNeIa) (Riess et al. 1998; Perlmutter et al. 1999) and Bennett et al. (2003) disclose that the Universe is currently accelerating. When these results combined with the observations of cosmic microwave background (Bennett et al. 2003; Spergel et al. 2003) and large scale structure (Tegmark et al. 2004a, b), strongly suggest that the Universe is spatially flat and dominated by an exotic component with large negative pressure called as dark energy (Weinberg 1989; Carroll 2001; Peebles and Ratra 2003; Padmanabhan 2003). Wilkinson Microwave Anisotropy Probe(WMAP) shows that dark energy occupies 73 % of the energy of our Universe. Dark matter occupies 23 % and rest 4 % energy is baryonic matter (ordinary matter) in the Universe. There are several candidates for dark energy: first is the cosmological constant (Padmanabhan 2003), and the second is the so-called dynamic candidates such as: Phantom (Caldwell et al.2003), quintessence (Capozziello et al. 2006), K-essence (Scherrer 2004), Chaplygin gas (Kamenshchik et al. 2001), quintom (Feng et al. 2006), holographic dark energy (Horava and Minic 2000), and extra dimensions (Rogatko 2004).

Recently, holographic dark energy models have received considerable attention to describe dark energy cosmological models. According to the holographic principle, the number of degrees of freedom in a bounded system should be finite and is related to the area of its boundary (Hooft 1993). Cosmological versions of holographic principle have been discussed by many authors Fischler and Susskind (1998), Tavakol and Ellis (1999), Easther and Lowe (1999). Chattopadhyay and Debnath (2009), Farajollahi et al. (2012), Debnath (2012), Malekjani (2013) are some of the authors who have investigated several aspects of holographic dark energy. Recently, Kiran et al. (2014, 2015) have studied minimally interacting dark energy models in scalar-tensor theories.

From last few decades there has been an immense interest in constructing cosmological models of the Universe to study the origin, physics and ultimate fate of the Universe. In particular, cosmological models of Brans-Dicke and Saez- Ballester scalar-tensor theories of gravitation are attracting more and more attention of scientists. A detailed discussion of Brans-Dicke and Saez-Ballester scalar-tensor theories is given in chapter-1. Survey of literature in Brans-Dicke theory is presented in chapter-5. Rao et al. (2007) have studied an exact Bianchi type-V cosmological model in Saez-Ballester theory of gravitation. Rao et al. (2008a) have discussed exact Bianchi type-II, VIII and IX string cosmological models in Saez-Ballester theory of gravitation. Rao et al. (2008b) have obtained exact Bianchi type-II, VIII and IX perfect fluid cosmological models in Saez-Ballester theory of gravitation. Rao et al. (2012c) have investigated a dark energy model in a scalar-tensor theory of gravitation. Naidu et al. (2012a) have discussed LRS Bianchi type-II dark energy model in a scalar-tensor theory of gravitation. Naidu et al. (2012b) have obtained LRS Bianchi type-*II* Universe with cosmic strings and bulk viscosity in a scalar tensor theory of gravitation. Rao et al. (2013c) have studied Bianchi type-*II*, *VIII* and *IX* perfect fluid and dark energy cosmological models in Saez-Ballester and general theory of gravitation. Recently, Rao et al. (2015f) have discussed five-dimensional bulk viscous cosmological model with wet dark fluid in Saez-Ballester theory of gravitation.

Inspired by the above investigations and discussion, in this chapter, we proposed Kantowski-Sachs cosmological model in Saez-Ballester theory of gravitation and Kaluza–Klein cosmological model in Brans-Dicke theory of gravitation with minimally interacting holographic dark energy.

This chapter is organized as follows: in section 7.2, we have obtained the Saez-Ballester field equations and their solution with the help of Kantowski-Sachs metric in the presence of matter and holographic dark energy. Also, some important features of the model with conclusions are discussed. Section 7.3 contains Brans-Dicke field equations and their solutions with the help of Kaluza-

Klein metric in the presence of matter and holographic dark energy. Also, some important features of the model with conclusions are discussed.

7.2 Kantowski-Sachs holographic Model in Saez-Ballester Theory

7.2.1 Metric and field equations:

We consider a spatially homogeneous Kantowski - Sachs metric of the form

$$ds^2 = dt^2 - A^2 dr^2 - B^2\left(d\theta^2 + Sin^2\theta\ d\varphi^2\right) \qquad (7.2.1)$$

where $A\ \&\ B$ are the functions of time t only.

The energy momentum tensors for holographic dark energy and dark matter (pressure less i.e. $w_m = 0$) are respectively given by

$$T_{ij} = (p_\Lambda + \rho_\Lambda)u_i u_j - p_\Lambda g_{ij} \qquad (7.2.2)$$

$$\overline{T}_{ij} = \rho_m u_i u_j \qquad (7.2.3)$$

Here ρ_m is the energy density of dark matter, ρ_Λ and p_Λ are the energy density and pressure of the holographic dark energy.

The field equations given by Saez-Ballester (1986) for the combined scalar and tensor fields are

$$R_{ij} - \frac{1}{2} R g_{ij} - \omega \phi^n \left(\phi_{,i} \phi_{,j} - \frac{1}{2} g_{ij} \phi_{,r} \phi^{,r} \right) = -8\pi T_{ij} \qquad (7.2.4)$$

and the scalar field ϕ satisfies the equation

$$2\phi^n \phi^{,i}_{;i} + n\phi^{n-1} \phi_{,r} \phi^{,r} = 0 \qquad (7.2.5)$$

where R is the scalar curvature, ω and n are constants, T_{ij} is the energy momentum tensor of the matter.

The energy conservation equation is

$$T^{ij}_{,j} = 0 \qquad (7.2.6)$$

In a commoving coordinate system, the Saez-Ballester field equations (7.2.4) and (7.2.5) for the metric (7.2.1) with the help of equations (7.2.2) and (7.2.3) can be written as

$$2\frac{\ddot{B}}{B} + \frac{\dot{B}^2}{B^2} + \frac{1}{B^2} + \frac{\omega}{2}\dot{\phi}^2 \phi^n = -p_\Lambda \qquad (7.2.7)$$

$$\frac{\ddot{A}}{A} + \frac{\ddot{B}}{B} + \frac{\dot{A}\dot{B}}{AB} + \frac{\omega}{2}\dot{\phi}^2 \phi^n = -p_\Lambda \qquad (7.2.8)$$

$$2\frac{\dot{A}\dot{B}}{AB} + \frac{\dot{B}^2}{B^2} + \frac{1}{B^2} - \frac{\omega}{2}\dot{\phi}^2 \phi^n = \rho_\Lambda + \rho_m \qquad (7.2.9)$$

$$\ddot{\phi} + \dot{\phi}\left(\frac{\dot{A}}{A} + 2\frac{\dot{B}}{B}\right) + \frac{n}{2}\frac{\phi^n}{\phi} = 0 \tag{7.2.10}$$

Also the energy conservation equation is given by

$$\dot{\rho}_\Lambda + \dot{\rho}_m + \left(\frac{\dot{A}}{A} + 2\frac{\dot{B}}{B}\right)(\rho_\Lambda + p_\Lambda + \rho_m) = 0 \tag{7.2.11}$$

where an overhead dot represents derivative with respect to time t. Here we are considering dark matter and holographic dark energy components are minimally interacting. Hence they conserve separately, so that we have

$$\dot{\rho}_m + \left(\frac{\dot{A}}{A} + 2\frac{\dot{B}}{B}\right)\rho_m = 0 \tag{7.2.12}$$

$$\dot{\rho}_\Lambda + \left(\frac{\dot{A}}{A} + 2\frac{\dot{B}}{B}\right)(1 + w_\Lambda)\rho_\Lambda = 0 \tag{7.2.13}$$

where $w_\Lambda = \dfrac{p_\Lambda}{\rho_\Lambda}$ is the equation of state parameter for holographic dark energy.

7.2.2 Solution of the field equations:

The equations (7.2.7) to (7.2.10) is a system of four independent equations with six unknowns $A, B, p_\Lambda, \rho_\Lambda, \rho_m$ and ϕ.

In order to get a deterministic solution we take the following possible physical condition, the shear scalar σ is proportional to scalar expansion θ, which leads to the following relationship

$$\text{i.e., } A = B^k \tag{7.2.14}$$

where k is an arbitrary constant.

From equations (7.2.7), (7.2.8) and (7.2.14),

we get

$$B = \left(k_1 t + k_2\right), \tag{7.2.15}$$

$$A = \left(k_1 t + k_2\right)^k, \tag{7.2.16}$$

where $k_1^2 = \dfrac{1}{k^2 - 1}$, $k \neq \pm 1$ and k_2 is an integration constant.

From equations (7.2.10), (7.3.15) and (7.3.16), we get the scalar field (ϕ) as

$$\phi = \left[\frac{k_3(n+2)}{2(k+1)}\left(k_1 t + k_2\right)^{-(k+1)}\right]^{\frac{2}{n+2}} \tag{7.2.17}$$

where k_3 is an integrating constant.

Now the metric (7.2.1) can be written as

$$ds^2 = dt^2 - \left(k_1 t + k_2\right)^{2k} dr^2 - \left(k_1 t + k_2\right)^2 \left(d\theta^2 + Sin^2\theta\, d\varphi^2\right)$$

$$\tag{7.2.18}$$

From equations (7.2.7), (7.2.8) and (7.2.15)-(7.2.17), we get

$$p_\Lambda = \frac{\omega k_3^{\,2}}{2(k_1 t + k_2)^{2k+4}} - \frac{(k^2+1)k_1^{\,2}+1}{2(k_1 t + k_2)^2} \qquad (7.2.19)$$

From equation (7.2.12) the energy density for dark matter as

$$\rho_m = \rho_0 (k_1 t + k_2)^{-(k+2)} \qquad (7.2.20)$$

where ρ_0 is an integrating constant.

Now from equations (7.2.9), (7.2.20) and (7.2.15)-(7.2.17), we get

$$\rho_\Lambda = \frac{(2k+1)k_1^{\,2}+1}{(k_1 t + k_2)^2} + \frac{\omega k_3^{\,2}}{2(k_1 t + k_2)^{2k+4}} - \rho_0 (k_1 t + k_2)^{-(k+2)} \qquad (7.2.21)$$

From (7.2.19) and (7.2.21), the holographic equation of state parameter w_Λ is given by

$$w_\Lambda = \frac{\omega k_3^{\,2} - \left((k^2+1)k_1^{\,2}+1\right)(k_1 t + k_2)^{2k+2}}{2\left((2k+1)k_1^{\,2}+1\right)(k_1 t + k_2)^{2k+2} + \omega k_3^{\,2} - 2\rho_0 (k_1 t + k_2)^{k+2}} \qquad (7.2.22)$$

Thus the metric (7.2.18) together with equations (7.2.19)-(7.2.22) constitutes Kantowski-Sachs minimally interacting holographic dark energy cosmological model in a scalar-tensor theory proposed by Saez and Ballester.

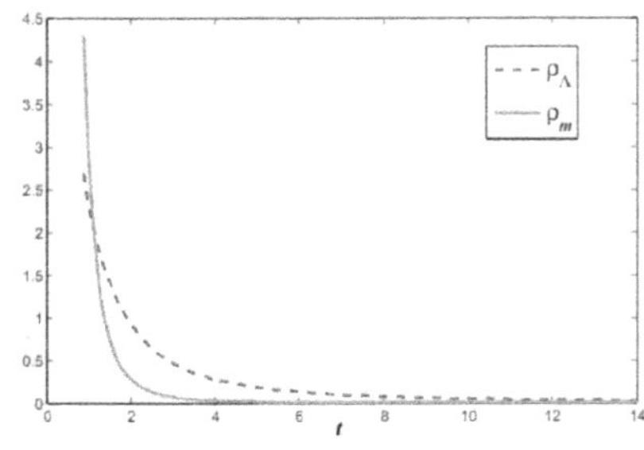

Fig. 7.1: Plot of energy density of holographic dark energy and dark matter versus t.

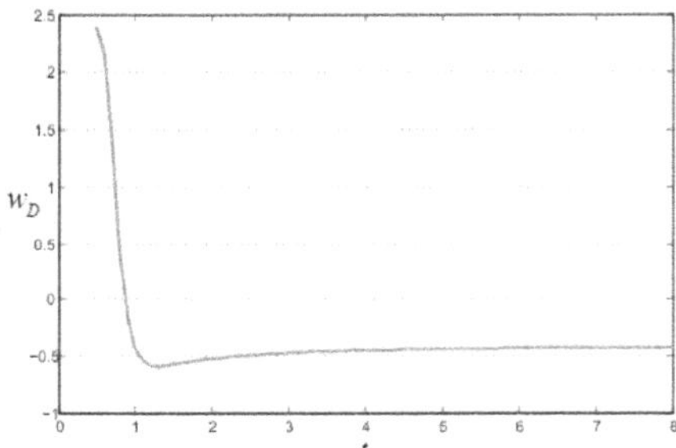

Fig. 7.2: Plot of EoS parameter of holographic dark energy versus t.

From figure-7.1, it is observed that the curve ρ_m which represents matter density dominates curve ρ_Λ which represents dark energy density in early stages of evolution of the Universe, whereas thereafter, the process is completely reverse and is observed that the dark energy density dominates the matter density.

177

Figure-7.2 describes the behavior of EoS parameter of holographic dark energy in terms of time t. We observed that the curve w_Λ varying from positive region to negative region with the increase of time t, which conforms the recent observations. Obtained model initially (in early stages of evolution of the Universe) dominated by dark matter and later on dominated by dark energy.

7.2.3 Some important features and conclusions:

The volume element and average scale factor of the model (7.2.18) are given by

$$V = \sqrt{-g} = \left(k_1 t + k_2\right)^{(k+2)} \sin\theta \tag{7.2.23}$$

$$a(t) = V^{1/3} = \left[\left(k_1 t + k_2\right)^{(k+2)} \sin\theta\right]^{\frac{1}{3}} \tag{7.2.24}$$

The expression for the expansion scalar θ is given by

$$\theta = u^i{}_{,i} = \frac{k_1(k+2)}{k_1 t + k_2} \tag{7.2.25}$$

The Hubble's parameter H is given by

$$H = \frac{k_1(k+2)}{3(k_1 t + k_2)} \tag{7.2.26}$$

and the shear σ is given by

$$\sigma^2 = \frac{1}{2}\sigma^{ij}\sigma_{ij} = \frac{7}{18}\left(\frac{k_1(k+2)}{k_1 t + k_2}\right)^2 \tag{7.2.27}$$

The mean anisotropy parameter A_m is given by

$$A_m = \frac{1}{3}\sum_{i=1}^{3}\left(\frac{H_i - H}{H}\right)^2 = 2\left(\frac{k-1}{k+2}\right)^2 \tag{7.2.28}$$

where $\Delta H_i = H_i - H$ $(i = 1,2,3)$

The overall density parameter Ω is given by

$$\Omega = \frac{\rho}{3H^2} = \frac{3}{2k_1^2(k+2)^2}\left(2(2k+1)k_1^2 + \omega k_3^2\left(k_1 t + k_2\right)^{-(2k+2)} + 2\right) \tag{7.2.29}$$

The deceleration parameter q is given by

$$q = -\frac{a\ddot{a}}{\dot{a}^2} = -1 + \frac{3}{k+2} \tag{7.2.30}$$

From equation (7.2.30), it is observed that deceleration parameter is within the range $-1 \leq q < 0$ for $k > 1$, hence the obtained model represent accelerated expansion of the Universe.

Jerk parameter

$$j = \frac{\dddot{a}}{aH^3} = \frac{(k-1)(k-4)}{(k+2)^2} \tag{7.2.31}$$

Look back time

$$\Delta t = (k+2)H_0^{-1}\left(1-(1+z)^{\frac{-3}{k+2}}\right) \qquad (7.2.32)$$

where H_0 is present value of Hubble's parameter and 'z' is red-

shift.

Luminosity distance $d_L = r_1 a_0 (1+z)$, where $r_1 = \int_t^{t_0} \frac{1}{a(t)} dt$

$$d_L = \frac{k(1+z)(k+2)}{(1-k)H_0}\left(1-(1+z)^{\frac{k-1}{k+2}}\right) \qquad (7.2.33)$$

The distance modulus $D(z)$ is given by

$$D(z) = 5\log d_L + 25$$

i.e. $\quad D(z) = 5\log\left[\frac{k(1+z)(k+2)}{(1-k)H_0}\left(1-(1+z)^{\frac{k-1}{k+2}}\right)\right] + 25. \qquad (7.2.34)$

Conclusions:

Here, we have obtained and presented spatially homogeneous Kantowski-Sachs cosmological model filled with matter and holographic dark energy components which are interacting minimally within the frame work of Saez-Ballester

(1986) scalar-tensor theory. The following are the observations and conclusions:

- The volume of the model vanishes at $t = t^*$ where $t^* = -k_2 / k_1$ and expansion scalar tends to infinity, which shows that the Universe starts evolving with zero volume at $t = t^*$ with an infinite rate of expansion. Also model has no singularity for $k > 0$.

- Our obtained model represents accelerated expansion of the Universe and also good agreement with the recent cosmological observations. Average anisotropic parameter $A_m \neq 0$ for $k \neq 1$, so this model is anisotropic throughout the evolution of the Universe.

- We have also obtained expressions for look back time, luminosity distance, distance modulus and jerk parameter.

7.3 Kaluza-Klein Holographic Model in Brans-Dicke Theory

7.3.1 Metric and field equations:

We consider spatially homogeneous five dimensional Kaluza–Klein metric in the form

$$ds^2 = dt^2 - A^2(t)(dx^2 + dy^2 + dz^2) - B^2(t)dm^2 \qquad (7.3.1)$$

where A and B are functions of 't' only .

Brans-Dicke (1961) theory of gravitation is a natural extension of general relativity which introduces an additional scalar field ϕ besides the metric tensor g_{ij} and dimensionless coupling constant ω . The Brans-Dicke field equations for combined scalar and tensor field are given by

$$R_{ij} - \frac{1}{2} R g_{ij} = -8\pi\phi^{-1} T_{ij} - \omega\phi^{-2}\left(\phi_{,i}\phi_{,j} - \frac{1}{2}g_{ij}\phi_{,r}\phi^{,r}\right)$$
$$\qquad\qquad\qquad\qquad - \phi^{-1}(\phi_{i;j} - g_{ij}\,\phi_{,r}^{\;,r}) \qquad (7.3.2)$$

and $\qquad \phi_{,r}^{\;,r} = 8\pi(3+2\omega)^{-1}T \qquad (7.3.3)$

where R is the scalar curvature, T_{ij} is the stress energy tensor of the matter, comma and semicolon denote partial and covariant differentiation respectively.

Also, we have energy conservation equation

$$T^{ij}_{;j} = 0 \tag{7.3.4}$$

This equation is a consequence of the field equations (7.3.2) and (7.3.3).

In co-moving coordinate systems, the Brans-Dicke field equations (7.3.2) and (7.3.3) for the metric (7.3.1) with the help of equations (7.2.2) and (7.2.3) can be written as

$$\frac{2\ddot{A}}{A} + \frac{\dot{A}^2}{A^2} + \frac{2\dot{A}\dot{B}}{AB} + \frac{\ddot{B}}{B} + \frac{\omega\dot{\phi}^2}{2\phi^2} + \frac{\ddot{\phi}}{\phi} + \left(\frac{2\dot{A}}{A} + \frac{\dot{B}}{B}\right)\frac{\dot{\phi}}{\phi} = -\frac{8\pi\, p_\Lambda}{\phi} \tag{7.3.5}$$

$$3\frac{\dot{A}^2}{A^2} + 3\frac{\dot{A}\dot{B}}{AB} - \frac{\omega\dot{\phi}^2}{2\phi^2} + \left(\frac{3\dot{A}}{A} + \frac{\dot{B}}{B}\right)\frac{\dot{\phi}}{\phi} = \frac{8\pi}{\phi}\left(\rho_\Lambda + \rho_m\right) \tag{7.3.6}$$

$$\frac{3\ddot{A}}{A} + \frac{3\dot{A}^2}{A^2} + \frac{\omega\dot{\phi}^2}{2\phi^2} + \frac{\ddot{\phi}}{\phi} + \frac{3\dot{A}\dot{\phi}}{A\phi} = -\frac{8\pi\, p_\Lambda}{\phi} \tag{7.3.7}$$

$$\ddot{\phi} + \dot{\phi}\left(\frac{3\dot{A}}{A} + \frac{\dot{B}}{B}\right) = \frac{8\pi}{(3+2\omega)}\left(\rho_\Lambda + \rho_m + p_\Lambda\right) \tag{7.3.8}$$

Also the energy conservation equation leads to

$$\left(\dot{\rho}_{\Lambda}+\dot{\rho}_{m}\right)+\left(\frac{3\dot{A}}{A}+\frac{\dot{B}}{B}\right)\left(\rho_{\Lambda}+\rho_{m}+p_{\Lambda}\right)=0 \qquad (7.3.9)$$

here the overhead dot denotes differentiation with respect to t.

Here we are considering the minimally interacting matter and holographic dark energy components. Hence both the components conserve separately, so that we have

$$\dot{\rho}_{m}+\left(\frac{3\dot{A}}{A}+\frac{\dot{B}}{B}\right)\rho_{m}=0 \qquad (7.3.10)$$

$$\dot{\rho}_{\Lambda}+\left(\frac{3\dot{A}}{A}+\frac{\dot{B}}{B}\right)(1+w_{\Lambda})\rho_{\Lambda}=0 \qquad (7.3.11)$$

where $w_{\Lambda}=\dfrac{p_{\Lambda}}{\rho_{\Lambda}}$ is the equation of state parameter for holographic dark energy.

7.3.2 Solution of the field equations:

The equations (7.3.5) to (7.3.8) is a system of four independent equations with six unknowns $A, B, p_{\Lambda}, \rho_{\Lambda}, \rho_{m}$ and ϕ. In order to get a deterministic solution we take the following plausible physical conditions,

i. The shear scalar σ is proportional to scalar expansion θ, which leads to the linear relationship between the metric potentials A and B,

$$\text{i.e., } B = A^n, \tag{7.3.12}$$

where n is an arbitrary constant.

ii. The relation between scalar filed ϕ and average scale factor $a(t)$ given by (Pimental 1985; Johri and Kalyani 1994)

$$\phi = \phi_0 a^k (t), \tag{7.3.13}$$

where ϕ_0 and k are arbitrary constants.

From equations (7.3.5), (7.3.7), (7.3.12) and (7.3.13), we get

$$A = \left(k_3 t + k_4\right)^{\frac{k_1}{k_3}}, \tag{7.3.14}$$

$$B = \left(k_3 t + k_4\right)^{\frac{n k_1}{k_3}}, \tag{7.3.15}$$

where $k_3 = \dfrac{(n+3)(k+4)k_1}{4}$, $k_4 = \dfrac{(n+3)(k+4)k_2}{4}$ and k_1, k_2 are integrating constants.

From equations (7.3.13) - (7.3.15), we get

$$\phi = \phi_0 \left(k_3 t + k_4\right)^{\frac{k}{k+4}} \tag{7.3.16}$$

Now the metric (7.3.1) can be written as

$$ds^2 = dt^2 - \left(k_3 t + k_4\right)^{\frac{2k_1}{k_3}}\left(dx^2 + dy^2 + dz^2\right) - \left(k_3 t + k_4\right)^{\frac{2nk_1}{k_3}} dm^2 \qquad (7.3.17)$$

From equations (7.3.5), (7.3.7) and (7.3.14) - (7.3.16), we get

$$p_\Lambda = \frac{\phi_0\left(k_3 t + k_4\right)^{\frac{-(k+8)}{k+4}}}{16\pi\left(k+4\right)^2}\left\{\begin{array}{c} kk_3{}^2\left(8 - \omega k\right) - k_1\left(k+4\right)^2\left[\left(n+5\right)\left(k_1 k_3\right) + k_1\left(n^2 + n + 4\right)\right] \\ + kk_1 k_3\left(k+4\right)\left(n+5\right) \end{array}\right\}$$

$$(7.3.18)$$

From the equation (7.3.10), (7.3.14) and (7.3.15), we get

$$\rho_m = \rho_0\left(k_3 t + k_4\right)^{\frac{-4}{k+4}} \qquad (7.3.19)$$

From equations (7.3.6), (7.3.14) - (7.3.16) and (7.3.19), we get

$$\rho_\Lambda = \frac{\phi_0\left(k_3 t + k_4\right)^{\frac{-(k+8)}{k+4}}}{16\pi\left(k+4\right)^2}\left\{6k_1{}^2\left(n+1\right)\left(k+4\right)^2 + kk_3\left(2k_1\left(n+3\right)\left(k+4\right) - \omega kk_3\right)\right\}$$

$$- \rho_0\left(k_3 t + k_4\right)^{\frac{4}{k+4}}$$

$$(7.3.20)$$

The holographic equation of state parameter is given by

$$w_\Lambda = \frac{\left\{kk_3{}^2\left(8 - \omega k\right) - k_1\left(k+4\right)^2\left[\left(n+5\right)\left(k_1 k_3\right) + k_1\left(n^2 + n + 4\right)\right] + kk_1 k_3\left(k+4\right)\left(n+5\right)\right\}}{\left\{6k_1{}^2\left(n+1\right)\left(k+4\right)^2 + kk_3\left(2k_1\left(n+3\right)\left(k+4\right) - \omega kk_3\right)\right\} - 16\pi\rho_0\phi_0{}^{-1}\left(k+4\right)^2\left(k_3 t + k_4\right)}$$

$$(7.3.21)$$

Thus the metric (7.3.17) together with equations (7.3.18)-(7.3.21)

constitutes Kaluza-Klein minimally interacting holographic dark energy cosmological model in a scalar tensor theory proposed by Brans and Dicke.

7.3.3 Some important features and conclusions:

The volume element and average scale factor of the model (7.3.17) are given by

$$V = \sqrt{-g} = \left(k_3 t + k_4\right)^{\frac{4}{k+4}} \tag{7.3.22}$$

$$a(t) = V^{\frac{1}{4}} = \left(k_3 t + k_4\right)^{\frac{1}{k+4}} \tag{7.3.23}$$

The expression for the expansion scalar θ is given by

$$\theta = u^i{}_{,i} = \frac{k_1(n+3)}{(k_3 t + k_4)} \tag{7.3.24}$$

and the shear σ is given by

$$\sigma^2 = \frac{1}{2}\sigma^{ij}\sigma_{ij} = \frac{7}{18}\left(\frac{k_1(n+3)}{(k_3 t + k_4)}\right)^2 \tag{7.3.25}$$

The Hubble's parameter H is given by

$$H = \frac{k_1(n+3)}{4(k_3 t + k_4)} \tag{7.3.26}$$

The mean anisotropy parameter A_m is given by

$$A_m = \frac{1}{4}\sum_{i=1}^{4}\left(\frac{H_i - H}{H}\right)^2 = 3\left(\frac{n-1}{n+3}\right)^2,$$
(7.3.27)

where $\Delta H_i = H_i - H \quad (i = 1,2,3)$

The overall density parameter Ω is given by

$$\Omega = \frac{\rho}{3H^2} =$$

$$\frac{\phi_0\left(k_3 t + k_4\right)^{\frac{-(3k+16)}{k+4}}}{48\pi k_1^2 (n+3)^2 (k+4)^2}\left\{6k_1^2(n+1)(k+4)^2 + kk_3\left(2k_1(n+3)(k+4) - \omega kk_3\right)\right\}$$

(7.3.28)

The deceleration parameter q is given by

$$q = -\frac{a\ddot{a}}{\dot{a}^2} = k + 3$$
(7.3.29)

The sign of q characterizes inflation of the Universe. A positive sign of q i.e. $q > 0$, correspond to decelerating model whereas negative sign of q (particularly $-1 \le q < 0$) indicates accelerating phase or inflationary model. From equation (7.3.29), it is observed that deceleration parameter is negative (i.e., $q < 0$) for $k < -3$, hence the obtained model represents accelerating Universe.

Jerk parameter

$$j = \frac{\ddot{a}}{aH^3} = (k+3)(2k+7) \tag{7.3.30}$$

Look back time

$$\Delta t = \frac{H_0^{-1}}{(k+4)}\left[1-(1+z)^{-(k+4)}\right], \tag{7.3.31}$$

where H_0 is present value of Hubble's parameter and 'z' is red-

shift.

Luminosity distance $d_L = r_1 a_0(1+z)$, where $r_1 = \int_t^{t_0} \frac{1}{a(t)}dt$

$$d_L = \frac{H_0^{-1}(1+z)}{(k+3)}\left[1-(1+z)^{-(k+3)}\right] \tag{7.3.32}$$

The distance modulus $D(Z) = 5\log d_L + 25$

i.e. $\quad D(Z) = 5\log\left\{\frac{H_0^{-1}(1+z)}{(k+3)}\left[1-(1+z)^{-(k+3)}\right]\right\} + 25 \tag{7.3.33}$

Conclusions:

In this chapter, we have obtained and presented spatially

homogeneous Kaluza-Klein cosmological model filled with two

minimally interacting fluids, matter and holographic dark energy

components in the frame work of Brans-Dicke scalar-tensor theory

(1961). The volume of the model vanishes at $t = t^*$, where $t^* = \dfrac{-k_4}{k_3}$ and expansion scalar is tends to infinity, which shows that the Universe starts evolving with zero volume at $t = t^*$ with an infinite rate of expansion. Also model has no singularity for $n > 0$. Our obtained model represents accelerated expansion of the Universe and also good agreement with the recent cosmological observations. Average anisotropic parameter $A_m \neq 0$ for $n \neq 1$, so this model is anisotropic. We have also obtained expressions for look back time, luminosity distance and jerk parameter. Thus the model presented here is anisotropic, non-rotating, shearing and also accelerating.

Chapter-8

Kantowski-Sachs dark energy model in general scalar-tensor theory[*]

* Work presented, in this chapter is covered by the following publication:

➢ Kantowski-Sachs Dark Energy Cosmological Model in General Scalar-Tensor Theory of Gravitation: Prespacetime Journal, 5, 1009 (2014).

8.1 Introduction

Nordtvedt (1970) proposed a general class of scalar-tensor gravitational theories in which the parameter ω of the BD theory is allowed to be an arbitrary function of the scalar field $[\omega \rightarrow \omega(\phi)]$. Based on Supernovae (SNe) type Ia observations, cosmologists have accepted the idea of dark energy, which is a fluid with negative pressure making up around 70% of the present Universe. Current studies to extract the properties of a dark energy component of the Universe from observational data focus on the determination of its equation of state $w(t)$, which is the ratio of the dark energy pressure to its energy density $w(t) = \dfrac{p}{\rho}$ and is not necessarily a constant. Recently, the parameter $w(t)$ has been calculated with some reasoning and the simplest dark energy candidate is the vacuum energy $(w = -1)$, which is mathematically equivalent to the cosmological constant (Λ). The other conventional alternatives, which can be described by minimally coupled scalar fields, are quintessence $(w > -1)$, phantom energy $(w < -1)$ and quintom (that can across from phantom region to quintessence region as evolved) and have time dependent EoS parameter. Ray et al. (2010), Yadav and Yadav (2010), Kumar (2010) and Pradhan et al. (2011) are some of the authors who have investigated dark energy models in general relativity with variable EoS parameter in different contexts. Rao and Sreedevi Kumari (2012) have discussed a cosmological model with negative constant deceleration parameter in general scalar-tensor theory of gravitation. Rao et al. (2012b) have obtained Kaluza-Klein radiating model in this theory. Rao and Neelima (2013d, e) have obtained LRS Bianchi type-*I* dark energy cosmological models in general scalar tensor theory of gravitation using two different expressions for $\omega(\phi)$. Rao and Neelima (2013f, g) have also obtained Bianchi types-*II, VIII* and *IX* and Kantowski-Sachs string with bulk viscous cosmological models respectively in this theory. Recently, Rao et al. (2015b) have discussed FRW bulk viscous cosmological model with wet dark fluid in general scalar-tensor theory.

Inspired by the above investigations and discussions, in this chapter, we focus our attention on investigating spatially homogeneous Kantowski - Sachs dark energy cosmological models in Nordtvedt (1970) general scalar tensor theory with the help of a special case proposed by Schwinger (1970), i.e. $3+2\omega(\phi)=1/\lambda\phi$, where λ is a constant.

The present chapter is organized as follows: In section 8.2, we established the field equations of general scalar-tensor theory proposed by Nordtvedt with the help of Kantowski-Sachs metric. We have obtained the solution of the filed equations in section 8.3. We discussed some important features of the model in section 8.4. The last section 8.5 contains some conclusions of the obtained model.

8.2 Metric and Field Equations

We consider a spatially homogeneous Kantowski-Sachs metric of the form

$$ds^2 = dt^2 - A^2 dr^2 - B^2 \left(d\theta^2 + \sin^2 \theta \, d\varphi^2 \right) \tag{8.2.1}$$

where A & B are the functions of time t only.

Beside the Bianchi type metrics, the Kantowski-Sachs (1966) models are also describing spatially homogeneous Universes. For a review of Kantowski-Sachs of metrics one can refer to MacCallum (1971). These metrics represent homogeneous but anisotropically expanding (or contacting) cosmologies and provide models where the effects of anisotropy can be estimated and compared with all well known Friedmann-Roberston-Walker class of cosmologies. Wang Xing-Xiang (2005) has obtained Kantowski - Sachs string cosmological model with bulk viscosity in general relativity. Kandalkar et al. (2009) have discussed Kantowski-Sachs viscous fluid cosmological model with a varying Λ. Kandalkar et al. (2011) have obtained string cosmology in Kantowski-Sachs space-time with bulk viscosity and magnetic field.

The energy momentum tensor components of the fluid can be written in anisotropic diagonal form as

$$T_j^i = diag\,[T_0^0, T_1^1, T_2^2, T_3^3]. \tag{8.2.2}$$

We can parameterize the components of the energy momentum tensor as follows:

$$\begin{aligned}
T_j^i &= diag[\rho, -p_x, -p_y, -p_z] \\
&= diag[1, -w_x, -w_y, -w_z]\rho \\
&= diag[1, -w, -(w+\gamma), -(w+\delta)]\rho
\end{aligned} \tag{8.2.3}$$

where ρ is the energy density of the fluid. p_x, p_y and p_z are the pressures, w_x, w_y and w_z are the directional equation of state (EoS) parameters of the fluid along x, y and z axes respectively and $w(t) = \dfrac{p}{\rho}$ is the deviation free EoS parameter of the fluid.

Here we have parameterized the deviation from isotropy by setting $w_x = w$. Also $\gamma \& \delta$ are the skewness parameters, which is a deviation from w along y and z axes. The parameters $w, \gamma \& \delta$ are not necessarily constants and can be functions of the cosmic time t.

Here u^i and x^i satisfy the equations

$$g_{ij}u^i u^j = 1, \; g_{ij}x^i x^j = -1, \quad \text{and} \quad u^i x_i = 0 \tag{8.2.4}$$

The field equations of general scalar-tensor theory proposed by Nordtvedt are

$$R_{ij} - \frac{1}{2}g_{ij}R = -8\pi\phi^{-1}T_{ij} - \omega\phi^{-2}\left(\phi_{,i}\,\phi_{,j} - \frac{1}{2}g_{ij}\phi_{,r}\,\phi^{,r}\right) - \phi^{-1}(\phi_{i;j} - g_{ij}\phi^{;r}_{;r})$$

$$(8.2.5)$$

$$\phi^{;r}_{;r} = \frac{8\pi T}{3+2\omega} - \frac{1}{(3+2\omega)}\frac{d\omega}{d\phi}\phi_{,i}\,\phi^{,i} \qquad (8.2.6)$$

where R_{ij} is the Ricci tensor, R is the scalar curvature, T_{ij} is the stress energy tensor of the matter, comma and semicolon denote partial and covariant differentiation respectively.

Also, we have the energy conservation equation

$$T^{ij}_{;j} = 0. \qquad (8.2.7)$$

8.3 Solution of the Field Equations

Using commoving coordinates, the field equations (8.2.5) to (8.2.7) for the metric (8.2.1) with the help of equations (8.2.2) to (8.2.4) can be written as

$$2\frac{\ddot{B}}{B}+\frac{\dot{B}^2}{B^2}+\frac{1}{B^2}+\frac{\omega\dot{\phi}^2}{2\phi^2}+\frac{\ddot{\phi}}{\phi}+2\frac{\dot{\phi}\dot{B}}{\phi B}=-\frac{8\pi}{\phi}w\rho \tag{8.3.1}$$

$$\frac{\ddot{A}}{A}+\frac{\ddot{B}}{B}+\frac{\dot{A}\dot{B}}{AB}+\frac{\omega\dot{\phi}^2}{2\phi^2}+\frac{\ddot{\phi}}{\phi}+\frac{\dot{\phi}}{\phi}\left(\frac{\dot{A}}{A}+\frac{\dot{B}}{B}\right)=-\frac{8\pi}{\phi}(w+\gamma)\rho \tag{8.3.2}$$

$$\frac{\ddot{A}}{A}+\frac{\ddot{B}}{B}+\frac{\dot{A}\dot{B}}{AB}+\frac{\omega\dot{\phi}^2}{2\phi^2}+\frac{\ddot{\phi}}{\phi}+\frac{\dot{\phi}}{\phi}\left(\frac{\dot{A}}{A}+\frac{\dot{B}}{B}\right)=-\frac{8\pi}{\phi}(w+\delta)\rho \tag{8.3.3}$$

$$2\frac{\dot{A}\dot{B}}{AB}+\frac{\dot{B}^2}{B^2}+\frac{1}{B^2}-\frac{\omega\dot{\phi}^2}{2\phi^2}+\frac{\dot{\phi}}{\phi}\left(\frac{\dot{A}}{A}+2\frac{\dot{B}}{B}\right)=\frac{8\pi}{\phi}\rho \tag{8.3.4}$$

$$\ddot{\phi}+\dot{\phi}\left(\frac{\dot{A}}{A}+2\frac{\dot{B}}{B}\right)=\frac{8\pi}{(3+2\omega)}(1-\gamma-\delta-3w)-\frac{1}{(3+2\omega)}\frac{d\omega}{d\phi}\dot{\phi}^2 \tag{8.3.5}$$

$$\dot{\rho}+\rho\left(\frac{\dot{A}}{A}+2\frac{\dot{B}}{B}\right)+w\rho\frac{\dot{A}}{A}+2(w+\gamma)\rho\frac{\dot{B}}{B}=0. \tag{8.3.6}$$

Here the over head dot denotes differentiation with respect to t.

From equations (8.3.2) and (8.3.3), we get

$$\gamma=\delta. \tag{8.3.7}$$

Using (8.3.7) and by using the transformation $dt=AB^2dT$, the above field equations (8.3.1) to (8.3.6) can be written as

$$2\frac{B''}{B} - 3\frac{B'^2}{B^2} - 2\frac{A'B'}{AB} + A^2B^2 + \frac{\omega}{2}\frac{\phi'^2}{\phi^2} + \frac{\phi''}{\phi} - \frac{\phi'}{\phi}\frac{A'}{A} = -\frac{8\pi}{\phi}w\rho\left(A^2B^4\right) \qquad (8.3.8)$$

$$\frac{A''}{A} - \frac{A'^2}{A^2} + \frac{B''}{B} - 2\frac{B'^2}{B^2} - 2\frac{A'B'}{AB} + \frac{\omega}{2}\frac{\phi'^2}{\phi^2} + \frac{\phi''}{\phi} - \frac{\phi'}{\phi}\frac{B'}{B} = \frac{-8\pi}{\phi}(w+\gamma)\rho\left(A^2B^4\right)$$

$$(8.3.9)$$

$$2\frac{A'B'}{AB} + \frac{B'^2}{B^2} + A^2B^2 - \frac{\omega}{2}\frac{\phi'^2}{\phi^2} + \frac{\phi'}{\phi}(\frac{A'}{A} + 2\frac{B'}{B}) = \frac{8\pi}{\phi}\rho\left(A^2B^4\right) \qquad (8.3.10)$$

$$(3+2\omega)\phi'' = 8\pi(1-2\gamma-3w)\rho(A^2B^4) - \phi'^2\frac{d\omega}{d\phi} \qquad (8.3.11)$$

$$\rho' + \rho\left(\frac{A'}{A} + 2\frac{B'}{B}\right) + w\rho\frac{A'}{A} + 2(w+\gamma)\rho\frac{B'}{B} = 0. \qquad (8.3.12)$$

From here after the over head dash denotes differentiation with respect to T.

The field equations (8.3.8) to (8.3.11) are four independent equations with seven unknowns $A, B, \omega, \phi, w, \gamma$ & ρ.

From equations (8.3.8) to (8.3.11), we have

$$(3+2\omega)\phi'' + \frac{d\omega}{d\phi}\phi'^2 = 2\phi\left[\frac{A''}{A} - \frac{A'^2}{A^2} + 2\frac{B''}{B} - 3\frac{B'^2}{B^2} - 2\frac{A'B'}{AB} + A^2B^2\right] + \omega\frac{\phi'^2}{\phi} + 3\phi''$$

$$(8.3.13)$$

Here we obtain dark energy cosmological model in Nordtvedt's general scalar-tensor theory with the help of a special case proposed by Schwinger (1970) in the form

$$3 + 2\omega(\phi) = \frac{1}{\lambda\phi}, \quad \lambda = \text{constant} \tag{8.3.14}$$

From equations (8.3.13) and (8.3.14), we get

$$\frac{1}{\lambda}\left[\frac{\phi''}{\phi} - \frac{\phi'^2}{\phi^2}\right] + \frac{3}{2}\frac{\phi'^2}{\phi} - 3\phi'' = 2\phi\left[\frac{A''}{A} - \frac{A'^2}{A^2} + 2\frac{B''}{B} - 3\frac{B'^2}{B^2} - 2\frac{A'B'}{AB} + A^2 B^2\right]$$

$$\tag{8.3.15}$$

In order to solve the above equation completely, we assume that the expansion scalar is proportional to shear scalar. This condition leads to

$$A = B^n, n \neq 0 \tag{8.3.16}$$

From equations (8.3.15) and (8.3.16), we get

$$\phi = e^{k_1 T + k_2} \tag{8.3.17}$$

where k_1 & k_2 are arbitrary constants and

$$\frac{B''}{B} - \frac{3(n+1)}{n+2}\frac{B'^2}{B^2} = \frac{-1}{(n+2)}B^{2n+2} - \frac{3}{4(n+2)}k_1^2 \tag{8.3.18}$$

From equation (8.3.18), we get

$$B = \left[\left(\frac{k_4}{k_3}\right) Sech(n+1)k_4 T\right]^{\frac{1}{n+1}}, n \neq -1 \qquad (8.3.19)$$

where $k_3^2 = \dfrac{1}{(n^2+n+1)}$ & $k_4^2 = \dfrac{3k_1^2}{4(2n+1)}$

From equations (8.3.16) and (8.3.19), we get

$$A = \left[\left(\frac{k_4}{k_3}\right) Sech(n+1)k_4 T\right]^{\frac{n}{n+1}} \qquad (8.3.20)$$

From equation (8.3.10), we get the string energy density

$$\rho = \frac{1}{8\pi k_5}\left[Cosh(n+1)k_4 T\right]^{\frac{2(n+2)}{n+1}} \left\{ e^{(k_1 T + k_2)} \left[\begin{array}{l} (2n+1)k_4^2 Tanh^2(n+1)k_4 T \\[6pt] + \dfrac{k_4^2}{k_3^2} Sech^2(n+1)k_4 T \\[6pt] -(n+2)k_1 k_4 Tanh(n+1)k_4 T + \frac{3}{4}k_1^2 \end{array} \right] - \frac{k_1^2}{4\lambda} \right\}$$

$$(8.3.21)$$

where $k_5 = \left(\dfrac{k_4}{k_3}\right)^{\frac{2(n+2)}{n+1}}$

From equation (8.3.8), we get the EoS parameter

$$w = -\frac{\left[\begin{array}{l}(2n+1)k_4^2 \operatorname{Tanh}^2((n+1)k_4 T) - (n^2+n+1)k_4^2 \operatorname{Sech}^2((n+1)k_4 T) \\[2mm] - nk_1 k_4 \operatorname{Tanh}((n+1)k_4 T) - \dfrac{k_1^2}{4}\end{array}\right] - \dfrac{k_1^2}{4\lambda} e^{-(k_1 T + k_2)}}{\left[\begin{array}{l}(2n+1)k_4^2 \operatorname{Tanh}^2((n+1)k_4 T) + \dfrac{k_4^2}{k_3^2} \operatorname{Sech}^2((n+1)k_4 T) \\[4mm] -(n+2)k_1 k_4 \operatorname{Tanh}((n+1)k_4 T) + \dfrac{3k_1^2}{4}\end{array}\right] - \dfrac{k_1^2}{4\lambda} e^{-(k_1 T + k_2)}}$$

$$(8.3.22)$$

From equations (8.3.8) and (8.3.9), we get the skewness parameter

$$\gamma = \frac{n(2n+1)k_4^2 \operatorname{Sech}^2((n+1)k_4 T) + (n-1)k_1 k_4 \operatorname{Tanh}((n+1)k_4 T)}{\left[\begin{array}{l}(2n+1)k_4^2 \operatorname{Tanh}^2((n+1)k_4 T) + \dfrac{k_4^2}{k_3^2} \operatorname{Sech}^2((n+1)k_4 T) \\[4mm] -(n+2)k_1 k_4 \operatorname{Tanh}((n+1)k_4 T) + \dfrac{3k_1^2}{4}\end{array}\right] - \dfrac{k_1^2}{4\lambda} e^{-(k_1 T + k_2)}}$$

$$(8.3.23)$$

201

The metric (8.2.1) can now be written as

$$ds^2 = \frac{1}{k_5}\left[Cosh(n+1)k_4 T\right]^{\frac{2(n+2)}{n+1}} dT^2 - -\left[\left(\frac{k_4}{k_3}\right)Sech(n+1)k_4 T\right]^{\frac{2n}{n+1}} dr^2$$

$$-\left[\left(\frac{k_4}{k_3}\right)Sech(n+1)k_4 T\right]^{\frac{2}{n+1}}\left(d\theta^2 + Sin^2\theta\, d\varphi^2\right)$$

$$(8.3.24)$$

The metric (8.3.24) together with (8.3.17) and (8.3.21) to (8.3.23) constitutes a Kantowski-Sachs anisotropic dark energy cosmological model in Nordtvedt's general scalar-tensor theory of gravitation with a special case proposed by Schwinger.

Isotropic cosmological model

For $n = 1$, the metric (8.3.24) together with (8.3.17) and (8.3.21) to (8.3.23) constitutes a Kantowski-Sachs isotropic dark energy cosmological model in Nordtvedt's general scalar-tensor theory of gravitation with a special case proposed by Schwinger.

8.4 Some Important Features of the Models

The volume element of the model (8.3.24) is given by

$$V = (-g)^{\frac{1}{2}} = \sqrt{k_5} \left[Sech(n+1)k_4 T \right]^{\frac{(n+2)}{n+1}} Sin\theta \tag{8.4.1}$$

The expression for the expansion scalar θ is given by

$$\theta = u^i{}_{;i} = -k_4(n+2)Tanh(n+1)k_4 T \tag{8.4.2}$$

and the shear σ is given by

$$\sigma^2 = \frac{1}{2}\sigma^{ij}\sigma_{ij} = \frac{7}{18}k_4{}^2(n+2)^2 Tanh^2(n+1)k_4 T \tag{8.4.3}$$

The deceleration parameter q is given by

$$q = -3\theta^{-2}\left(\theta_{,i}u^i + \frac{1}{3}\theta^2\right)$$

$$= -1 + \frac{6n}{(n+2)\left(\cos 2h(n+1)k_4 T - 1\right)} \tag{8.4.4}$$

The Hubble parameter H is given by

$$H = \frac{-k_4(n+2)Tanh(n+1)k_4 T}{3} \tag{8.4.5}$$

The tensor of rotation $w_{ij} = u_{i,j} - u_{j,i}$ is identically zero and hence this Universe is non-rotational.

8.5 Conclusions

In this chapter, we have presented Kantowski-Sachs dark energy cosmological models in Nordtvedt (1970) general scalar tensor theory with the help of a special case proposed by Schwinger (1970). The models presented here are free from singularities and the spatial volume vanishes as $T \to \infty$. We observe that, as T approaches to infinity, the expansion scalar θ leads to a constant value. This shows that the Universe expands homogeneously. We also observe that the shear scalar σ and the Hubble parameter H approaches to constant value as $T \to \infty$. We can also observe that the energy density ρ diverges as $T \to \infty$. The EoS parameter w, skewness parameter γ approach constant value as $T \to \infty$. For this model, the deceleration parameter $q = -1$ for large values of T and hence the Universe expands exponentially. The models obtained here remain anisotropic throughout the evolution of the Universe for $n \neq 1$. Study of anisotropies connects and unravels fundamental issues in various fields of astrophysics and cosmology. Anisotropic features can reveal key information on the structure and the nature of the components of the Universe, and provide hints on the origin of high energy emission. Hence anisotropic space-times are important. But for $n=1$, the model reduces to isotropic Universe. These anisotropic as well as isotropic exact models are free from singularities. Also the models presented here represent not only the early stages of evolution but also the present Universe.

Kantowski-Sachs dark energy model in general scalar-tensor theory*

*** Work presented, in this chapter is covered by the following publication:**

➢ **Kantowski-Sachs Dark Energy Cosmological Model in General Scalar-Tensor Theory of Gravitation: Prespacetime Journal, 5, 1009 (2014).**

8.1 Introduction

Nordtvedt (1970) proposed a general class of scalar-tensor gravitational theories in which the parameter ω of the BD theory is allowed to be an arbitrary function of the scalar field $[\omega \rightarrow \omega(\phi)]$. Based on Supernovae (SNe) type Ia observations, cosmologists have accepted the idea of dark energy, which is a fluid with negative pressure making up around 70% of the present Universe. Current studies to extract the properties of a dark energy component of the Universe from observational data focus on the determination of its equation of state $w(t)$, which is the ratio of the dark energy pressure to its energy density $w(t) = \dfrac{p}{\rho}$ and is not necessarily a constant. Recently, the parameter $w(t)$ has been calculated with some reasoning and the simplest dark energy candidate is the vacuum energy $(w = -1)$, which is mathematically equivalent to the cosmological constant (Λ). The other conventional alternatives, which can be described by minimally coupled scalar fields, are quintessence $(w > -1)$, phantom energy $(w < -1)$ and quintom (that can across from phantom region to quintessence region as evolved) and have time dependent EoS parameter. Ray et al. (2010), Yadav and Yadav (2010), Kumar (2010) and Pradhan et al. (2011) are some of the authors who have investigated dark energy models in general relativity with variable EoS parameter in different contexts. Rao and Sreedevi Kumari (2012) have discussed a cosmological model with negative constant deceleration parameter in general scalar-tensor theory of gravitation. Rao et al. (2012b) have obtained Kaluza-Klein radiating model in this theory. Rao and Neelima (2013d, e) have obtained LRS Bianchi type-I dark energy cosmological models in general scalar tensor theory of gravitation using two different expressions for $\omega(\phi)$. Rao and Neelima (2013f, g) have also obtained Bianchi types-II, VIII and IX and Kantowski-Sachs string with bulk viscous cosmological models respectively in this theory. Recently, Rao et al. (2015b) have discussed FRW bulk viscous cosmological model with wet dark fluid in general scalar-tensor theory.

Inspired by the above investigations and discussions, in this chapter, we focus our attention on investigating spatially homogeneous Kantowski - Sachs dark energy cosmological models in Nordtvedt (1970) general scalar tensor theory with the help of a special case proposed by Schwinger (1970), i.e. $3 + 2\omega(\phi) = 1/\lambda\phi$, where λ is a constant.

The present chapter is organized as follows: In section 8.2, we established the field equations of general scalar-tensor theory proposed by Nordtvedt with the help of Kantowski-Sachs metric. We have obtained the solution of the filed equations in section 8.3. We discussed some important features of the model in section 8.4. The last section 8.5 contains some conclusions of the obtained model.

8.2 Metric and Field Equations

We consider a spatially homogeneous Kantowski-Sachs metric of the form

$$ds^2 = dt^2 - A^2 dr^2 - B^2 \left(d\theta^2 + \sin^2 \theta\, d\varphi^2 \right) \qquad (8.2.1)$$

where A & B are the functions of time t only.

Beside the Bianchi type metrics, the Kantowski-Sachs (1966) models are also describing spatially homogeneous Universes. For a review of Kantowski-Sachs of metrics one can refer to MacCallum (1971). These metrics represent homogeneous but anisotropically expanding (or contacting) cosmologies and provide models where the effects of anisotropy can be estimated and compared with all well known Friedmann-Roberston-Walker class of cosmologies. Wang Xing-Xiang (2005) has obtained Kantowski - Sachs string cosmological model with bulk viscosity in general relativity. Kandalkar et al. (2009) have discussed Kantowski-Sachs viscous fluid cosmological model with a varying Λ. Kandalkar et al. (2011) have obtained string cosmology in Kantowski-Sachs space-time with bulk viscosity and magnetic field.

The energy momentum tensor components of the fluid can be written in anisotropic diagonal form as

$$T_j^i = diag\,[T_0^0, T_1^1, T_2^2, T_3^3].$$

(8.2.2)

194

We can parameterize the components of the energy momentum tensor as follows:

$$
\begin{aligned}
T^i_j &= diag[\rho,-p_x,-p_y,-p_z] \\
&= diag[1,-w_x,-w_y,-w_z]\rho \\
&= diag[1,-w,-(w+\gamma),-(w+\delta)]\rho
\end{aligned}
\qquad (8.2.3)
$$

where ρ is the energy density of the fluid. p_x, p_y and p_z are the pressures, w_x, w_y and w_z are the directional equation of state (EoS) parameters of the fluid along x, y and z axes respectively and $w(t) = \dfrac{p}{\rho}$ is the deviation free EoS parameter of the fluid.

Here we have parameterized the deviation from isotropy by setting $w_x = w$. Also $\gamma\,\&\,\delta$ are the skewness parameters, which is a deviation from w along y and z axes. The parameters $w, \gamma\,\&\,\delta$ are not necessarily constants and can be functions of the cosmic time t.

Here u^i and x^i satisfy the equations

$$
g_{ij}u^i u^j = 1,\; g_{ij}x^i x^j = -1, \quad \text{and} \quad u^i x_i = 0 \qquad (8.2.4)
$$

The field equations of general scalar-tensor theory proposed by Nordtvedt are

$$R_{ij} - \frac{1}{2} g_{ij} R = -8\pi \phi^{-1} T_{ij} - \omega \phi^{-2} \left(\phi,_i \, \phi,_j - \frac{1}{2} g_{ij} \phi,_r \, \phi^{,r} \right) - \phi^{-1} (\phi_{i;j} - g_{ij} \phi^{;r}_{;r})$$

$$(8.2.5)$$

$$\phi^{;r}_{;r} = \frac{8\pi T}{3+2\omega} - \frac{1}{(3+2\omega)} \frac{d\omega}{d\phi} \phi,_i \, \phi^{,i}$$ (8.2.6)

where R_{ij} is the Ricci tensor, R is the scalar curvature, T_{ij} is the stress energy tensor of the matter, comma and semicolon denote partial and covariant differentiation respectively.

Also, we have the energy conservation equation

$$T^{ij}_{;j} = 0.$$ (8.2.7)

8.3 Solution of the Field Equations

Using commoving coordinates, the field equations (8.2.5) to (8.2.7) for the metric (8.2.1) with the help of equations (8.2.2) to (8.2.4) can be written as

$$2\frac{\ddot{B}}{B}+\frac{\dot{B}^2}{B^2}+\frac{1}{B^2}+\frac{\omega\dot{\phi}^2}{2\phi^2}+\frac{\ddot{\phi}}{\phi}+2\frac{\dot{\phi}\dot{B}}{\phi B}=-\frac{8\pi}{\phi}w\rho \qquad (8.3.1)$$

$$\frac{\ddot{A}}{A}+\frac{\ddot{B}}{B}+\frac{\dot{A}\dot{B}}{AB}+\frac{\omega\dot{\phi}^2}{2\phi^2}+\frac{\ddot{\phi}}{\phi}+\frac{\dot{\phi}}{\phi}\left(\frac{\dot{A}}{A}+\frac{\dot{B}}{B}\right)=-\frac{8\pi}{\phi}(w+\gamma)\rho \qquad (8.3.2)$$

$$\frac{\ddot{A}}{A}+\frac{\ddot{B}}{B}+\frac{\dot{A}\dot{B}}{AB}+\frac{\omega\dot{\phi}^2}{2\phi^2}+\frac{\ddot{\phi}}{\phi}+\frac{\dot{\phi}}{\phi}\left(\frac{\dot{A}}{A}+\frac{\dot{B}}{B}\right)=-\frac{8\pi}{\phi}(w+\delta)\rho \qquad (8.3.3)$$

$$2\frac{\dot{A}\dot{B}}{AB}+\frac{\dot{B}^2}{B^2}+\frac{1}{B^2}-\frac{\omega\dot{\phi}^2}{2\phi^2}+\frac{\dot{\phi}}{\phi}\left(\frac{\dot{A}}{A}+2\frac{\dot{B}}{B}\right)=\frac{8\pi\,\rho}{\phi} \qquad (8.3.4)$$

$$\ddot{\phi}+\dot{\phi}\left(\frac{\dot{A}}{A}+2\frac{\dot{B}}{B}\right)=\frac{8\pi}{(3+2\omega)}(1-\gamma-\delta-3w)-\frac{1}{(3+2\omega)}\frac{d\omega}{d\phi}\dot{\phi}^2 \qquad (8.3.5)$$

$$\dot{\rho}+\rho\left(\frac{\dot{A}}{A}+2\frac{\dot{B}}{B}\right)+w\rho\frac{\dot{A}}{A}+2(w+\gamma)\rho\frac{\dot{B}}{B}=0. \qquad (8.3.6)$$

Here the over head dot denotes differentiation with respect to t.

From equations (8.3.2) and (8.3.3), we get

$$\gamma=\delta. \qquad (8.3.7)$$

Using (8.3.7) and by using the transformation $dt=AB^2dT$, the above field equations (8.3.1) to (8.3.6) can be written as

$$2\frac{B''}{B} - 3\frac{B'^2}{B^2} - 2\frac{A'B'}{AB} + A^2 B^2 + \frac{\omega}{2}\frac{\phi'^2}{\phi^2} + \frac{\phi''}{\phi} - \frac{\phi'}{\phi}\frac{A'}{A} = -\frac{8\pi}{\phi}w\rho\left(A^2 B^4\right) \qquad (8.3.8)$$

$$\frac{A''}{A} - \frac{A'^2}{A^2} + \frac{B''}{B} - 2\frac{B'^2}{B^2} - 2\frac{A'B'}{AB} + \frac{\omega}{2}\frac{\phi'^2}{\phi^2} + \frac{\phi''}{\phi} - \frac{\phi'}{\phi}\frac{B'}{B} = \frac{-8\pi}{\phi}(w+\gamma)\rho\left(A^2 B^4\right)$$

$$(8.3.9)$$

$$2\frac{A'B'}{AB} + \frac{B'^2}{B^2} + A^2 B^2 - \frac{\omega}{2}\frac{\phi'^2}{\phi^2} + \frac{\phi'}{\phi}(\frac{A'}{A} + 2\frac{B'}{B}) = \frac{8\pi}{\phi}\rho\left(A^2 B^4\right) \qquad (8.3.10)$$

$$(3+2\omega)\phi'' = 8\pi(1-2\gamma-3w)\rho(A^2 B^4) - \phi'^2\frac{d\omega}{d\phi} \qquad (8.3.11)$$

$$\rho' + \rho\left(\frac{A'}{A} + 2\frac{B'}{B}\right) + w\rho\frac{A'}{A} + 2(w+\gamma)\rho\frac{B'}{B} = 0. \qquad (8.3.12)$$

From here after the over head dash denotes differentiation with respect to T.

The field equations (8.3.8) to (8.3.11) are four independent equations with seven unknowns $A, B, \omega, \phi, w, \gamma$ & ρ.

From equations (8.3.8) to (8.3.11), we have

$$(3+2\omega)\phi'' + \frac{d\omega}{d\phi}\phi'^2 = 2\phi\left[\frac{A''}{A} - \frac{A'^2}{A^2} + 2\frac{B''}{B} - 3\frac{B'^2}{B^2} - 2\frac{A'B'}{AB} + A^2 B^2\right] + \omega\frac{\phi'^2}{\phi} + 3\phi''$$

$$(8.3.13)$$

Here we obtain dark energy cosmological model in Nordtvedt's general scalar-tensor theory with the help of a special case proposed by Schwinger (1970) in the form

$$3 + 2\omega(\phi) = \frac{1}{\lambda\phi}, \quad \lambda = \text{constant} \tag{8.3.14}$$

From equations (8.3.13) and (8.3.14), we get

$$\frac{1}{\lambda}\left[\frac{\phi''}{\phi} - \frac{\phi'^2}{\phi^2}\right] + \frac{3}{2}\frac{\phi'^2}{\phi} - 3\phi'' = 2\phi\left[\frac{A''}{A} - \frac{A'^2}{A^2} + 2\frac{B''}{B} - 3\frac{B'^2}{B^2} - 2\frac{A'B'}{AB} + A^2B^2\right]$$

$$\tag{8.3.15}$$

In order to solve the above equation completely, we assume that the expansion scalar is proportional to shear scalar. This condition leads to

$$A = B^n, n \neq 0 \tag{8.3.16}$$

From equations (8.3.15) and (8.3.16), we get

$$\phi = e^{k_1 T + k_2} \tag{8.3.17}$$

where k_1 & k_2 are arbitrary constants and

$$\frac{B''}{B} - \frac{3(n+1)}{n+2}\frac{B'^2}{B^2} = \frac{-1}{(n+2)}B^{2n+2} - \frac{3}{4(n+2)}k_1^2 \tag{8.3.18}$$

From equation (8.3.18), we get

$$B = \left[\left(\frac{k_4}{k_3}\right) Sech(n+1)k_4T\right]^{\frac{1}{n+1}}, n \neq -1 \tag{8.3.19}$$

where $k_3^{\;2} = \dfrac{1}{(n^2+n+1)}$ & $k_4^{\;2} = \dfrac{3k_1^{\;2}}{4(2n+1)}$

From equations (8.3.16) and (8.3.19), we get

$$A = \left[\left(\frac{k_4}{k_3}\right) Sech(n+1)k_4T\right]^{\frac{n}{n+1}} \tag{8.3.20}$$

From equation (8.3.10), we get the string energy density

$$\rho = \frac{1}{8\pi k_5}\left[Cosh(n+1)k_4T\right]^{\frac{2(n+2)}{n+1}}\left\{e^{(k_1T+k_2)}\left[\begin{array}{l}(2n+1)k_4^{\;2}Tanh^2(n+1)k_4T\\[4pt]+\dfrac{k_4^{\;2}}{k_3^{\;2}}Sech^2(n+1)k_4T\\[6pt]-(n+2)k_1k_4Tanh(n+1)k_4T+\dfrac{3}{4}k_1^{\;2}\end{array}\right]-\dfrac{k_1^{\;2}}{4\lambda}\right\} \tag{8.3.21}$$

where $k_5 = \left(\dfrac{k_4}{k_3}\right)^{\frac{2(n+2)}{n+1}}$

From equation (8.3.8), we get the EoS parameter

$$w = -\frac{\left[\begin{array}{l}(2n+1)k_4^2\,Tanh^2(n+1)k_4 T - (n^2+n+1)k_4^2\,Sech^2(n+1)k_4 T \\[2mm] -\,nk_1 k_4\,Tanh(n+1)k_4 T - \dfrac{k_1^2}{4}\end{array}\right]-\dfrac{k_1^2}{4\lambda}e^{-(k_1 T + k_2)}}{\left[\begin{array}{l}(2n+1)k_4^2\,Tanh^2(n+1)k_4 T + \dfrac{k_4^2}{k_3^2}\,Sech^2(n+1)k_4 T \\[2mm] -\,(n+2)k_1 k_4\,Tanh(n+1)k_4 T + \dfrac{3k_1^2}{4}\end{array}\right]-\dfrac{k_1^2}{4\lambda}e^{-(k_1 T + k_2)}}$$

(8.3.22)

From equations (8.3.8) and (8.3.9), we get the skewness parameter

$$\gamma = \frac{n(2n+1)k_4^2\,Sech^2(n+1)k_4 T + (n-1)k_1 k_4\,Tanh(n+1)k_4 T}{\left[\begin{array}{l}(2n+1)k_4^2\,Tanh^2(n+1)k_4 T + \dfrac{k_4^2}{k_3^2}\,Sech^2(n+1)k_4 T \\[2mm] -\,(n+2)k_1 k_4\,Tanh(n+1)k_4 T + \dfrac{3k_1^2}{4}\end{array}\right]-\dfrac{k_1^2}{4\lambda}e^{-(k_1 T + k_2)}}$$

(8.3.23)

The metric (8.2.1) can now be written as

$$ds^2 = \frac{1}{k_5}\left[Cosh(n+1)k_4 T\right]^{\frac{2(n+2)}{n+1}} dT^2 - -\left[\left(\frac{k_4}{k_3}\right)Sech(n+1)k_4 T\right]^{\frac{2n}{n+1}} dr^2$$

$$-\left[\left(\frac{k_4}{k_3}\right)Sech(n+1)k_4 T\right]^{\frac{2}{n+1}}\left(d\theta^2 + Sin^2\theta\, d\varphi^2\right)$$

$$(8.3.24)$$

The metric (8.3.24) together with (8.3.17) and (8.3.21) to (8.3.23) constitutes a Kantowski-Sachs anisotropic dark energy cosmological model in Nordtvedt's general scalar-tensor theory of gravitation with a special case proposed by Schwinger.

Isotropic cosmological model

For $n = 1$, the metric (8.3.24) together with (8.3.17) and (8.3.21) to (8.3.23) constitutes a Kantowski-Sachs isotropic dark energy cosmological model in Nordtvedt's general scalar-tensor theory of gravitation with a special case proposed by Schwinger.

8.4 Some Important Features of the Models

The volume element of the model (8.3.24) is given by

$$V = (-g)^{\frac{1}{2}} = \sqrt{k_5} \left[Sech(n+1)k_4 T \right]^{\frac{(n+2)}{n+1}} Sin\theta \tag{8.4.1}$$

The expression for the expansion scalar θ is given by

$$\theta = u^i{}_{;i} = -k_4(n+2)Tanh(n+1)k_4 T \tag{8.4.2}$$

and the shear σ is given by

$$\sigma^2 = \frac{1}{2}\sigma^{ij}\sigma_{ij} = \frac{7}{18}k_4{}^2(n+2)^2 Tanh^2(n+1)k_4 T \tag{8.4.3}$$

The deceleration parameter q is given by

$$q = -3\theta^{-2}\left(\theta_{,i}u^i + \frac{1}{3}\theta^2\right)$$

$$= -1 + \frac{6n}{(n+2)(\cos 2h(n+1)k_4 T - 1)} \tag{8.4.4}$$

The Hubble parameter H is given by

$$H = \frac{-k_4(n+2)Tanh(n+1)k_4 T}{3} \tag{8.4.5}$$

The tensor of rotation $w_{ij} = u_{i,j} - u_{j,i}$ is identically zero and hence this Universe is non-rotational.

8.5 Conclusions

In this chapter, we have presented Kantowski-Sachs dark energy cosmological models in Nordtvedt (1970) general scalar tensor theory with the help of a special case proposed by Schwinger (1970). The models presented here are free from singularities and the spatial volume vanishes as $T \to \infty$. We observe that, as T approaches to infinity, the expansion scalar θ leads to a constant value. This shows that the Universe expands homogeneously. We also observe that the shear scalar σ and the Hubble parameter H approaches to constant value as $T \to \infty$. We can also observe that the energy density ρ diverges as $T \to \infty$. The EoS parameter w, skewness parameter γ approach constant value as $T \to \infty$. For this model, the deceleration parameter $q = -1$ for large values of T and hence the Universe expands exponentially. The models obtained here remain anisotropic throughout the evolution of the Universe for $n \neq 1$. Study of anisotropies connects and unravels fundamental issues in various fields of astrophysics and cosmology. Anisotropic features can reveal key information on the structure and the nature of the components of the Universe, and provide hints on the origin of high energy emission. Hence anisotropic space-times are important. But for $n=1$, the model reduces to isotropic Universe. These anisotropic as well as isotropic exact models are free from singularities. Also the models presented here represent not only the early stages of evolution but also the present Universe.

CPSIA information can be obtained
at www.ICGtesting.com
Printed in the USA
BVHW070846300123
657303BV00009B/1098